100 Questions (and Answers) About Research Methods

To my ever-questioning and wonderful friend, Russ Shafer-Landau

100 Questions (and Answers) About Research Methods

Neil J. Salkind
University of Kansas

Los Angeles | London | New Delhi
Singapore | Washington DC

KH

Los Angeles | London | New Delhi
Singapore | Washington DC

FOR INFORMATION:

SAGE Publications, Inc.
2455 Teller Road
Thousand Oaks, California 91320
E-mail: order@sagepub.com

SAGE Publications Ltd.
1 Oliver's Yard
55 City Road
London, EC1Y 1SP
United Kingdom

SAGE Publications India Pvt. Ltd.
B 1/I 1 Mohan Cooperative Industrial Area
Mathura Road, New Delhi 110 044
India

SAGE Publications Asia-Pacific Pte. Ltd.
33 Pekin Street #02–01
Far East Square
Singapore 048763

Acquisitions Editor: Vicki Knight
Associate Editor: Lauren Habib
Editorial Assistant: Kalie Koscielak
Production Editor: Libby Larson
Copy Editor: Liann Lech
Typesetter: C&M Digitals (P) Ltd.
Proofreader: Joyce Li
Indexer: Terri Corry
Cover Designer: Candice Harman
Marketing Manager: Helen Salmon
Permissions Editor: Adele Hutchinson

Printed in the United States of America

Library of Congress Cataloging-in-Publication Data

Salkind, Neil J.

100 questions (and answers) about research methods / Neil J. Salkind.

p. cm.
Includes index.

ISBN 978-1-4129-9203-9 (pbk.)

1. Social sciences—Research—Methodology. I. Title. II. Title: One hundred questions (and answers) about research methods.

H62.S3195 2012 001.4′2—dc22 2011002803

This book is printed on acid-free paper.

11 12 13 14 15 10 9 8 7 6 5 4 3 2 1

8/23/12

Contents

Preface

In an increasingly data-driven world, it is more important than ever for students as well as professionals to better understand the process of research, from the initial asking of questions through the analysis and interpretation of data leading to a final report, and everything in between.

100 Questions (and Answers) About Research Methods is an attempt to summarize the most important questions that lay in those in-between spaces that one could ask about research methods while providing an answer as well.

It was written because I have noticed in my own years of teaching that many students and other professionals need a short review or a book to steer them in the right direction as to what topics they need to focus on and where they might look to find further information.

This is a short book and intended for those individuals who need a refresher as to what the important topics are within this area of study as well as those who are entirely new to the discipline and need a resource as to what the key questions are. Think of *100 Questions* . . . as a reminder, a resource and refresher of sorts. It's for graduate students preparing for comprehensive exams, researchers who need a reference, undergraduates in affiliated programs who will not be taking a primary course in research methods. and anyone curious about how these tools can be used most effectively.

Some tips about using the book . . .

1. The questions are divided into nine parts as follows:
   - Part 1: Understanding the Research Process and Getting Started
   - Part 2: Reviewing and Writing About Your Research Question
   - Part 3: Introductory Ideas About Ethics
   - Part 4: Research Methods: Knowing the Language, Knowing the Ideas
   - Part 5: Sampling Ideas and Issues
   - Part 6: Describing Data Using Descriptive Techniques
   - Part 7: All About Testing and Measuring
   - Part 8: Understanding Different Research Methods
   - Part 9: All About Inference and Significance

2. Each of these questions and answers can stand alone and provide a direct question and relatively short answer. Probably a book-length response could be written for each question, but the task at hand here is quick—enough information to allow the reader to gain some knowledge before he or she moves on to the next question or topic.

3. Not all of these questions and answers are independent of one another; most supplement each other. This is to help ensure that important material is reinforced and to help make sure that primary and secondary topics get consideration as well.

4. Each question ends with reference to three other questions related to the current question. These are the three that I think can best supplement the primary question being asked and answered. There are, of course, many others.

Acknowledgments

Thanks to the best editorial and production teams in the business, especially to Vicki Knight, who enthusiastically greeted this idea and saw it through to contract and then book. She tolerated an unbelievable number of late-night emails from me and did it with grace and support and great assistance. To associate editor Lauren Habib, who continues to make things happen quickly and efficiently and has been a huge help. Thanks also to Liann Lech (best copy editor in the galaxy); Libby Larson, production editor; Kalie Koscielak, editorial assistant; and Helen Salmon, executive marketing manager. I own all the errors here, and please let me know via email how this book might be better.

And to the best kids any parents could wish for and have: Sara, Micah, and Ted; and of course, Lucky, Pepper's heir apparent; and Lew M., indefatigable, caring, courageous, and honest friend.

Neil J. Salkind
Lawrence, Kansas
njs@ku.edu

About the Author

Neil J. Salkind is Professor Emeritus at the University of Kansas, where he taught in the Department of Educational Psychology for more than 35 years. His interest is in writing about statistics and research methods in an informative, nonintimidating, and noncondescending way. He is also the author of *Statistics for People Who (Think They) Hate Statistics, Statistics for People Who (Think They) Hate Statistics (the Excel Edition), Excel Statistics—a Quick Guide,* and the recently edited *Encyclopedia of Research Design.*

PART 1

UNDERSTANDING THE RESEARCH PROCESS AND GETTING STARTED

QUESTION #1

Why Is Research Necessary, and What Are Some of Its Benefits?

This is a terrific question to start with and one to which every beginning (or refreshing) student should know the answer.

Today, more than ever, we live in a world that is driven by data and in which there is an increasing dependence on what the research process is all about, how it is carried out, and what the benefits (and drawbacks) might be.

Research is the complex process through which ideas are explored—new ones that yield unique information about some phenomenon of interest as well as already-established ideas that continue to raise important questions about outcomes that we find to be of importance.

In your own field, you can easily think of some of the many questions that are yet unanswered and that, through the process of research, will generate new and important findings.

For example, for the school psychologist, having some knowledge about the relationship between the maturity of the brain and adolescents' emotions may be helpful in understanding classroom behavior. Or, for the nursing student, it's understanding the effectiveness of caregiver groups in helping make caregivers more effective and better communicators with their charges. Or, it's the classroom teacher who wants to know a better way to teach children who are more visual than aural in their learning skills.

It's situations like these, and thousands of other opportunities, that have led researchers to develop tools and techniques that provide the basis for answers—and if not answers, then enough information help make informed decisions, and that is one of the keys.

So, why research? Here are just some of the benefits . . .

- It provides us with a basis for making decisions.
- As best as possible, it ensures that our decisions are based on data and not arbitrary or personal bias.
- If done correctly, it matches the approach of other scientists.
- It works.

More questions? See #5, #9, and #11.

QUESTION #2

Generally, What Is the Process Through Which a Research Topic Is Identified, and Do I Have to Be an Experienced Researcher to Select a Topic of Interest to Me?

The process varies for different researchers and different topics, but it basically can be any one of the following or some combination. But, regardless of where you get your initial ideas, we are assuming that there is something in which you are interested, regardless of your level of knowledge or previous experience.

If you are a beginner and this is your first effort at conducting research, then you might want to consult some of the many general sources, such as periodicals, newspapers, and websites, as examples for information about a topic you find of interest. At this point, what you want to do is find out what generally has been done in the area, what the important questions are, and what might be some of the most important considerations to take into account.

If you are more experienced, then you could seek out an expert in the field. That person may be your academic adviser or another faculty member who you know has some knowledge of the area of interest. Always feel comfortable asking these experts questions and seeking out their advice as to what has been done and what needs to be done in the future.

And, if you are really experienced in a particular area, you can go right to the most immediate sources of information about a particular area of interest, which are the journals and monographs on certain topics. You can access these in hard copy at the library, but of course, these days you can also access them online.

If you pursue a particular topic as your research area, you will probably end up with the last suggestion above, but it never can hurt to use all these strategies to get a comprehensive view of all the dimensions that make up a topic worthy of further study.

More questions? See #12, #17, and #19.

QUESTION #3

What Is the "Scientific Method," and How Can I Apply That to My Own Research?

Although there are more contemporary views of the scientific method, its start can be found with the Greek philosopher Aristotle, who lived some 2,300 years ago. He believed that reasoning should be the basis for all decision making, and this foundation has grown into a method of investigation that is defined as a set of techniques for collecting observable data and subjecting that data to experimentation.

In practical terms, the scientific method is used to test the effectiveness or value of several alternative methods or treatments. The method consists of two or more treatments (often called experimental groups and often including a control group). These groups are exposed to different conditions, and then the outcomes of some measure are compared to one another to determine if the groups performed differently. In theory, because the groups are equal on everything but their exposure to some condition, any difference observed would most likely be due to the treatment itself.

For example, every fall, millions of parents scramble to get their children enrolled in supplementary programs that purport to increase SAT test scores with the hope of ensuring entrance to the college of their choice. Do these extra (and expensive) programs work? One way to tell would be to compare different groups of participants that receive different levels of instruction, with another group of participants that receives no extra instruction at all. We assume that the groups are equal (and under the scientific method, there are many ways to test this assumption), and we can test to see which of the groups improved most or scored highest (or whatever measure is used).

In this example, experimentation on the performance of several groups is used to collect empirical observations (the test scores) to see if the program had an impact. If done correctly, the scientific method lends a very strong argument for making sound judgments about the relationship between ideas and a fair and sensible test of those ideas.

More questions? See #8, #27, and #32.

QUESTION #4

There Are Different Types of Research Models That One Might Use. Can You Give Me a General Overview and How They Might Compare?

There are two general categories of research models: nonexperimental and experimental.

Nonexperimental models, in general, are those in which there is no active manipulation of variables or no treatment taking place. Within this general category, you will find the following . . .

- Historical models (examining events that have occurred in the past)
- Descriptive models (describing things as they occur)
- Correlational models (looking at the relationship between variables)
- Qualitative models (examining outcomes and the social context within which they occur).

Experimental models, in general, are those in which there is active manipulation of variables and treatments or varying conditions are tested. Within this general category, you will find the following . . .

- Quasi-experimental (looking at the effects of a treatment in which one set of participants is preassigned to groups)
- Experimental (looking at the effects of a treatment in which participants are randomly assigned to groups)

Here's an example of each of these different methods within these two large categories.

	Model	Example
Non-experimental	Historical	Examining the occurrence of child discipline practices in colonial America and comparing them to current practices.
	Descriptive	Surveying college students to find out the number of hours of sleep they get and whether they feel it has an impact on their success in school.
	Correlational	Looking at the relationship between social media involvement and number of significant friends.
	Qualitative	Investigating the success of a charter school and its impact on urban and rural families.
Experimental	Quasi-Experimental	Examining differences in compliance levels among diabetic and nondiabetic adults in use of weight reduction programs.
	Experimental	Examining differences among three different types of balance programs that enroll older (80+ years) senior citizens.

More questions? See #3, #5, and #11.

QUESTION #5

What Is the Best Research Model for My Purposes?

Let's get right to the point. And, as if you didn't already suspect this, the model you follow depends upon the question you are asking.

The best model for your research is the one that most accurately answers the question you are asking. Here are a few things to keep in mind . . .

1. No one model or research method is the perfect fit for answering a question. Sometimes, a correlational model might fit quite nicely, yet there are other parts of a research question that might be better answered using other techniques.
2. Often, mixed models are the best approach because a question requires all the available tools you have at your disposal. For example, you may want to examine attitudes toward public assistance as a function of experience, and that would be done through a type of cross-sectional study. But, you may also be interested in the phenomenon of how the attitudes of particular families who are receiving public assistance have changed over time as well, which would call for a longitudinal approach.
3. Questions build upon previous work, and few questions stand alone. Be sure that your question has a strong foundation before even considering how you will answer it.

Given these three caveats, selecting a research model is pretty straightforward.

- Interested in describing only outcomes, such as using a survey to determine how many people prefer what kind of services offered by city government? Then you would look to a descriptive model.
- Interested in past events and how they relate to current events, such as how nurses in the public health service used to deal with infectious diseases and how they do now? Then the historical model might be best.

- Interested in looking at the relationship between variables, such as seeing if there is an association between income levels and attitude toward charter schools? Choose a correlation model.
- Interested in knowing whether a certain factor has a cause-and-effect outcome on another factor, such as the effect of weight training on cognitive skills in older Americans? You would use an experimental model here.

More questions? See #4, #11, and #14.

QUESTION #6

What Is the Difference Between Basic and Applied Research?

Quite a bit, and there are several ways to define these terms and distinguish between them.

One distinction has to do with the focus of the research. Basic research usually focuses on a very small part of a specific question or subject area under investigation. For example, if we were interested in better understanding the process through which nerve bundles in the brain become entangled as part of studying Alzheimer's disease, that would be basic research. Basic research is research that does not have any immediate application, whereas applied research does have immediate application.

On the other hand, a more far-reaching use of the results of this basic research might be in understanding how changes in exercise routine or diet can prevent such entanglement and thereby reduce the incidence of Alzheimer's disease This could be considered "more" applied. In other words, how immediately the results of such research can be used in real-world problems becomes one of the defining factors.

Another way to look at the difference is by examining what drives the research effort. For basic research, there is no other goal than the accumulation of knowledge and a better understanding of some phenomenon. Basic research is the foundation for all research, and it is sometimes difficult to foresee its benefits. Conducting basic research is the first step in moving toward applied research.

In the same vein of thinking, applied research attempts to solve problems that are more practical in nature, and, if we can be so bold, to help provide solutions, based on the results of basic research, that help others. Understanding why malnutrition affects brain growth is a basic and important question; understanding the implications of being malnourished on brain development and subsequent learning and success in school is important as well, but very definitely more applied.

More questions? See #1, #4, and #11.

QUESTION #7

What Is Qualitative Research, and What Are Some Examples?

Qualitative research examines individuals, institutions, and phenomena within the context in which they occur. It differs in many ways from quantitative research, but increasingly, there are aspects of each that blend into the other.

The qualitative researcher is interested in gaining an in-depth understanding of behavior and the reasons for that behavior. Qualitative research is relatively new to the social and behavioral sciences, having a much longer history in business and even medicine.

Whereas a strict quantitative researcher would look to empirical data that are quantitative in nature (test scores, etc.), a qualitative researcher might look to different sources for information, such as archival records, emails, open-ended surveys, videos, physical artifacts, direct observation, transcripts, participant observation and interviews, and focus groups. And, it is not so much the sources of the information that are important, but how they are used to answer the research question.

For example, case studies are an often-used method by qualitative researchers. If the questions asked have to do with how an institution ensures that its primary mission is accomplished, the information that is collected and the way that it is analyzed can differ greatly between qualitative methods and other methods.

Examples of types of studies that are conducted within a qualitative framework are case studies, in which individuals or institutions are examined in great depth, and ethnographic research projects, in which cultures are studied along with the customs that have developed within those cultures. And just as with quantitative research, there is a host of computer programs that can help aggregate and analyze qualitative information. Among the most popular are NVivo, Xsight, and Ethnograph.

More questions? See #4, #5, and #17.

QUESTION #8

What Are Hypotheses, and How Do They Fit Into the Scientific Method?

The word *hypothesis* comes from the Greek combination "to put under" or "to suppose," and it is not difficult to understand how we use the word to mean an "educated guess" or "a hunch." Hypotheses come from those theories on which scientists focus, and as a tool, a hypothesis helps us organize the information we are studying and drives our activities.

Hypotheses come in different flavors of the null and research variety, and they have very distinct purposes, but how do they fit into the scientific method?

The very essence of the scientific method is collecting empirical or observable data and then applying a rigorous (and public) testing of a clear question (often reflected in a hypothesis). Within the experimental model, this question most often includes a comparison between a control group and one or more experimental groups.

One of the major assumptions of this model is that both the control and the experimental group are basically the same in every way except for the fact that the experimental group receives some type of treatment. You should be aware that sometimes a control group is identified and may also receive treatment. Whether this is the case depends upon the question being asked. Is one interested in differences between treatment versus no treatment, or in treatment of different kinds regardless of the condition in which no treatment is applied?

The control group acts as the standard against which the results of the experiment (and those of the experimental group) are compared, and using some straightforward statistical tools, any observed differences are considered as to whether they are consistent or whether they occurred by chance.

More questions? See #3, #28, and #33.

QUESTION #9

What Do Good Research Hypotheses Do?

Research hypotheses are the "if-then" statements that are an important part of the scientific method. You may remember that ideas generate hypotheses to be tested, and that the results of these tests circle back to have an impact on how a theory might be modified.

Getting hypotheses right is very important, and here are some suggestions that you might find helpful.

First, a research hypothesis is always stated in the declarative. For example, *There is a negative relationship between number of injuries and amount of time spent warming up for high school track athletes.* Questions such as "What is the relationship . . ." aren't declarative statements and are less direct and less useful.

Second, a research hypothesis has to be testable, such as, *Children in families who participate in reading programs will score higher on standardized reading tests than those children whose families do not participate.* Here, the variables are clearly defined (participation in the reading program, or not) and scores on some standardized outcome. And for it to be testable, a hypothesis also has to be reasonable to test (you can find participants, the variables are easily measured, etc.).

Third, a research hypothesis sets out to test an outcome and not to prove it. For example, if a researcher is interested in the effects of social participation in seniors on longevity, the research hypothesis could be, *Participation in a twice-weekly social group for adults aged 80 and above results in an increased sense of belonging and an overall increase in life satisfaction as measured by the XYZ test.* This hypothesis is testable and one that does need proving or not.

Fourth, a good research hypothesis (and this might be as important as any of the other characteristics) has to be *based on theory or an idea worth testing.* A test of such a hypothesis allows researchers to add more knowledge to important questions (posed as theories) and help make the next step in the research process even more promising.

Finally, hypotheses have to be clearly stated and replicable, use variables that are possible to measure, and be able to be tested in an appropriate amount of time.

More questions? See #8, #29, and #30.

QUESTION #10

Besides Looking at the Reputation of a Journal Where a Study Is Published as One Criterion for a Good Study, Are There Other Things That I Can Look To?

Absolutely. Here's a checklist of a bunch of different criteria organized by categories that you can use to evaluate whether the study does what you think it should. Even though you have to make some subjective judgments here, check to see if you believe that the criterion is met, almost met, or not met at all.

		Is the criterion met?		
Category	**Criterion**	**Yes**	**Maybe**	**No**
The Review of Previous Research	Is the review of literature recent? Does the review hit the major and important points?	______	______	______
The Problem and the Purpose	Is the purpose of the study stated clearly?	______	______	______
	Is there a rationale for why the study is an important one to do?	______	______	______
	Are the research hypotheses clearly stated?	______	______	______

		Is the criterion met?		
Category	**Criterion**	**Yes**	**Maybe**	**No**
	Are the hypotheses grounded in theory or in a review and presentation of relevant literature?	______	______	______
The Method	Are the definitions and descriptions of the variables complete?	______	______	______
The Sample	Is it clear where the sample comes from and how it was selected?	______	______	______
Results and Discussion	Are the results related to the hypothesis?			
References	Are the references complete?	______	______	______

More questions? See #9, #12, and #14.

QUESTION #11

I Hear So Much About Different Studies—From the Newspaper, From Professional Bulletins, and Even From My Boss. What Am I Supposed to Believe, and How Can I Judge if the Results of a Study Are Useful?

This is really a good question and a tough one to answer.

First, we're going to assume that some research studies are done "better" than others—the researchers are more careful, they take the necessary time, use valid and reliable tools to measure outcomes, and don't go far beyond what the data allow them to conclude.

So you pick up your daily newspaper and on the front page is an article detailing the results of a study that shows that reading to children is an effective way to increase their vocabulary.

You teach young children and find these findings quite interesting. What do you do to judge whether these are really useful and applicable to your classroom setting? Let's suggest the following . . .

1. Find out where the original journal article on which the newspaper piece is based was published. The gold standard of whether research is "good" and conducted properly is the system where it is reviewed by the author's peers. Once the original article on which the newspaper article is based was completed in draft form, it was sent off to a journal editor, who then sent it out to experts in the field to have it reviewed. These experts read through the article carefully and came back with a recommendation to

   a. Publish the article (now called a manuscript) without change
   b. Make certain changes before it is published

c. Reject the article but invite the author(s) to resubmit it because the reviewers think it has some promise, or
d. Reject the article and don't invite a resubmission. Rejection rates run from about 85% for the best journals to about 15% for the least respected, and this is true for most fields of study.

2. Take a look at the journal (and you can do this for most online) and look at the editorial board. Do they represent major research universities? Schools that have a good reputation? These are the people reviewing the manuscript and are responsible for whether the research results eventually get published.

3. Ask your professor (if you are in a master's or higher program) or your specialist in your school district if they know of the journal or the people who did the work. Often, the largest body of work in a particular area is done by the same people or the same group of people at a few different institutions.

4. Finally, Google the authors of the research to see what other work they have done and how their work has been reviewed. You can learn a lot about the value of a particular research study by seeing what reviewers and other scholars have to say about it and the work of the people who authored it.

The bottom line on judging the value of research is to look at it from as many different perspectives as possible, including the author's credentials, where the work is published, and what other people (those you may respect for their knowledge of the field) think about it.

There is a huge amount of information published every day, and to make the most of it as it applies to your work, you have to be somewhat diligent and judge.

More questions? See #4, #5, and #15.

QUESTION #12

What Are Some of the Best Ways to Find Information Online, and Where Are Some of the Best Places?

You've probably heard this many times before in your career, but the library is almost always the best place to start (and certainly online), and the library's holdings can really provide you with a good start on the information you need.

The first step in building up your research skills and finding out about online resources is to get connected to your local library, both the public one as well as the community college or 4-year college or university. In most locations, if you are a resident of a community (or state), then you can get a library card and have access to everything the library has online, which is thousands of documents from everyday newspapers to journal articles to complete book collections.

What the library connection offers you is immediate access to publications, usually with no charge. Although it is always a good idea to visit the library and learn how to use the physical facilities, the electronic offerings are invaluable. And, in many cases, you can order a hard copy of a journal article or a copy of a book to be made available if you want to use it the old-fashioned way.

Once you are online and connected to your local library, many also offer live "chat" help, where you can talk with a librarian and ask such questions as "How do I search the *New York Times* cumulative index?" Also, virtual tours are usually offered online for the introduction you may need.

Other places to begin with are Google for general inquiries, which you may already use in your everyday searching activity. But less well known is Google Books at books.google.com, where you can search through categories such as medicine, humor, and political science, and preview copies of books in the Google collection. Also, there is Google Scholar, where you can search for scholarly books by the author or title, as well as journal articles and other types of publications.

No matter what search tool you use, look for options that allow you to save these references, or even tools that help you generate a bibliography in the appropriate format without having to retype information.

More questions? See #14, #17, and #19.

QUESTION #13

What Role Might Social Media Play in My Efforts as Both a Researcher and a Consumer?

You're probably very familiar with social media as tools for communication, and there's no question that such tools can be used productively in the research process.

Most obvious is the feature that allows one to follow other researchers and their activities through Twitter, Facebook, and LinkedIn (the latter mostly used by the business world, but there's no reason that the research community cannot use it as well).

Here's a very brief review of some social media tools and how they can be used.

Facebook has more than 500 million users, and anyone can create a Facebook page and deem it as his or her research center. Here, groups of colleagues can virtually meet, ideas can be exchanged, and the general topic advertised throughout the Facebook kingdom; other Facebook participants can even participate in studies as well. The average Facebook user is connected to about 80 other users, so this gives you some idea how vast the potential is for sharing ideas.

Twitter's more than 75 million users create 140-character messages to send out to those following them, or search for other Twitter users who are interested in a particular content. You can, of course, follow a particular researcher's work and contact that person for additional information—the most straightforward use of Twitter. You can also enter search terms using the # prefix (such as #infant) into Twitter's simple search tool on the main page and get the connections you want and perhaps some of the references you need.

Facebook, Twitter, and LinkedIn are just the social media tools that have become most popular. Others, such as Digg and Reddit, provide news, views, important coverage of current events, and information about the changing nature of society—all possibly important sources of information for the new researcher.

More questions? See #12, #17, and #19.

PART 2

REVIEWING AND WRITING ABOUT YOUR RESEARCH QUESTION

QUESTION #14

What Is a Review of the Literature, and Why Is It Important?

The review of literature is a summary of the important and relevant topics related to the topic you are studying. It usually comes after the introduction in a research proposal or a research report and is as long as it takes to introduce the reader to the perspectives about the topic under discussion. It should be comprehensive and succinct, and it should flow easily so the reader fully understands the past nature of the topic and the nature of the current research. It usually ends with a statement of the research question and the specific hypotheses that are to be explored.

The review of literature is a historical account of the research and thinking that has been done in the past as well as a review of the most current developments. It's a record of what has come before, and in the review, you are expected to summarize this record. You will also discuss the ideas relevant to what you are studying as well as the methods and the conclusions that were reached in previous studies—all in the service of your own interests so that your work is as current and relevant to the field as possible.

But even more important, the review of literature is a foundation upon which your work is based. From the review, which will consist mostly of primary and secondary sources, you will gain some direction regarding what your exact research question is and what hypotheses will be generated (and then tested) based on these questions.

Another reason why the review is so important is that it gives you the opportunity to better adjust your specific research question so that it reflects those important previous studies and the future direction they suggest. It's the opportunity to fine-tune your research question and hypotheses so they reflect exactly what you think is important and what it is you want to know.

More questions? See #15, #16, and #18.

QUESTION #15

How Does a Review of the Literature Have an Impact on My Research Question and the Hypothesis I Propose?

As you remember, a review of the literature is a process that is shaped like a funnel, where lots and lots of information goes into the wide top and out comes a very refined and specific research question at the more narrow bottom that is then tested via one or more hypotheses.

Along the way, you will proceed from a simple inquiry that you may have about something that interests you to even more interest about the specifics of a topic and then on to a more clearly stated research question. However, as you continue to read current and past research, and explore what has been done in the past and what might be done in the future, there is a significant give-and-take between what is known and what you want to further know. This can be illustrated by the back and forth between the formulation of the research question and the final writing process.

1. General interest in an idea

↓

2. Reading of general, primary, and secondary resources

↓

3. Tracking of important sources and documenting them (here's where the writing begins)

↓

4. Formulation of research question(s)

↓

5. Continued reading and organizing of important resources

↓

6. Formulation of hypothesis(es)

↓

7. Continued documentation and writing of literature review

More questions? See #14, #16, and #27.

QUESTION #16

How Do I Know When My Literature Review Is Finished? Couldn't It Go on Forever?

So you've reached the end of your review—or at least you think you have—and you really want to know when your review of literature is done. Here are some signs along the way (and, we are assuming that your research question is answerable and your hypothesis is testable).

You're done when . . .

1. All the references in the bibliography of newly discovered journal articles, chapters, and other primary and secondary sources have already been explored. This means that you have come full circle and what you are now looking at (be it a new or earlier reference) has already been consulted or referred to along the way.

2. Same thing with ideas. You have (more or less) exhausted the collection of sources about your research question and hypothesis, and although there's always something new to learn, for this exercise, you have come as far as you can.

3. Most important, you and your adviser and colleagues (and whoever else is acting as your mentor) are satisfied that you have comprehensively addressed the important issues of the theory underlying the question you are asking as well as the methods. This is a somewhat subjective criterion but very important nonetheless.

You're not done if . . .

1. You continue to find new (general) themes for the basic idea and important references and sources that you have never seen.

2. Your research question is suggested by authors in many other completed reports, chapters, and journal articles, and you should assume that someone else has addressed these suggestions. Look harder.

3. Most important, you just can't seem to finish the review of the literature or come up with additional ideas and such. Probably, your question is too ambitious (it may just be too grand in scope), and your hypothesis might not even be testable.

More questions? See #14, #17, and #18.

QUESTION #17

What Are the Three Main Sources of Information, and What Part Does Each Play in Creating a Literature Review?

There are three sources of information that you should consult as you write the review of literature: general, primary, and secondary. Each plays a different role and will provide you with different levels and types of information.

General sources (such as periodicals, newspapers, and news websites) are the most comprehensive and "friendly" sources of information about a particular topic. They are the best starting point when you want a general introduction to a topic and perhaps to find out the big names in the field who you can then contact directly. If health care and tax policy is what you want to study, for example, then a stop at the *New York Times* Index (at your library) or the *New York Times* Article Archive (online) are examples of a good first step to help you refine the question you want answered.

Secondary sources are those that are once removed from the original research and consist of reviews of research, anthologies of writings on a particular topic, or online discussions about specific issues, among other sources. These secondary sources are more focused reviews of the important issues surrounding a particular topic and are more in-depth than general sources. Often, finding one comprehensive review of a particular topic can be very helpful in refining a research question.

Finally, primary sources are those that report original research, and these reports most often take the form of articles in journals such as *Child Development, Educational and Psychological Measurement, The American Journal of Nursing,* and *Family Practice.* Your community or school library card is your key to accessing these first-person accounts of research studies and outcomes.

More questions? See #12, #14, and #16.

QUESTION #18

What Steps Should I Take in Writing My Review of Literature?

The review of research or review of literature sets the stage for the data you are to collect, the way it will be analyzed, and the conclusions that you reach. You want it to be comprehensive but also limited to your research question and hypothesis.

Here are some steps you might want to consider in writing your review of literature.

1. However you choose to do it, become very familiar with your online library resources or the actual bricks-and-mortar physical facility. This may mean taking an online workshop or a tour of the library and getting to know where the reference librarians are and when they are available for help.
2. Read colleagues' reviews of literature in your own area of interest to see how they organized their work and how it leads from a somewhat general orientation to the topic to a very refined hypothesis.
3. By this time, you should have a good idea of your research question. Now start seeking out general, secondary, and primary resources. Read whatever you can as you refine your question, and summarize those readings as you move along.
4. Read and summarize material that is highly relevant to your topic, and be sure as you work that you (always) track the sources and where you got the information for the bibliography that will accompany your proposal (and your final report).
5. As you read, generate a three-level table of contents.
6. Now, organize the summaries of the research you have been reading using the table of contents as the framework.

Once you have the table of contents organized and the various summaries of research within different levels of that table of contents, you are ready to take these notes and transform them into a written narrative. As you work, see to it that your colleague or adviser or teacher gets a chance to comment on your writing.

More questions? See #12, #15, and #16.

QUESTION #19

What Are Some of the Best Electronic Resources Available, and How Do I Learn to Use Them?

Before I make any suggestions regarding which electronic resources might be a very good place to start and, generally, how to use them, you must first have access to these resources. If you do not already have access, visit your university or college library (the building or online), and if you are not a student, then your local community library can provide access. If you do not have a computer with an Internet connection, the public libraries have them available as well.

Which electronic resource you use depends a great deal on what type of information you want to access and what your field of study is. For the purpose here, we're listing five general and five primary and secondary resources. But, these are lists of five out of many that are available, and your reference librarian can help you better choose.

Because each resource tends be different, you will have to explore how to use the specific tool. However, most have help, and you can also consult with a reference librarian on- and offline when needed. Keep in mind that the URLs you see here may be similar to, but not the same as, those you access through your library.

General Sources	URL
New York Times Index	http://www.nytimes.com/ref/membercenter/nytarchive.html
Time Magazine Archives	http://www.time.com/time/archive/
United States Newspapers	http://www.50states.com/news/
Reader's Guide to Periodical Literature	http://www.hwwilson.com/databases/readersg.cfm

Primary and Secondary Sources	
ERIC	http://www.hwwilson.com/databases/readersg.cfm
Social Sciences Index	http://www.hwwilson.com/dd/ss_i.htm
Medline	http://www.ncbi.nlm.nih.gov/pubmed/
PsycInfo	http://www.apa.org/pubs/databases/psycinfo/index.aspx
Expanded Academic Index	http://www.gale.cengage.com/PeriodicalSolutions/academicAsap.htm?grid=ExpandedAcademicASAPRedirect

A few more tips . . .

1. Knowing how to form search terms is the critical skill to learn. "School reform" yields very different results from "reform schools."
2. The more precise and targeted your search, the more accurate and useful will be the results.
3. Every search engine you use (be it Google, Chrome, or Bing, for example) has specific search algorithms, so use operators (such as "+" and "or") to narrow or expand your search. And, they each have a Help menu that can guide you in using these well.
4. You may find electronic resources to be all you need, but there's nothing like a trip to the library to see what you missed if you are looking only online.

More questions? See #12, #14, and #17.

INTRODUCTORY IDEAS ABOUT ETHICS

QUESTION #20

What Are Some of the More General and Important Principles of Ethical Research?

Almost every professional organization has a code by which it expects its members (and, in fact, members of the profession it represents) to abide. Here are some of the guidelines that many of these organizations hold in common.

- The first, and most important, priority is that no harm will come to those who participate, and this includes psychological, emotional, and physical harm.
- Research proposals should be reviewed by a group of people that represents the larger interests of the institution to which the researcher belongs. This review group is usually called something like an Institutional Review Board.
- All participants must sign, and understand, an informed consent form.
- Research that involves children and participants from special populations are to be reviewed using a different set of criteria to ensure that their rights are protected and they aren't exposed to dangerous or threatening situations. This is the case given that they are not usually the ones to agree to participate (their parents or guardians are), so they have little direct knowledge of the circumstances.
- If deception is involved as part of the research, then all participants must be debriefed following the experiment or the session in which the deception took place.
- Conflicts of interest must constantly be addressed and either removed from the research protocol or taken into account so they do not pose a danger to the participant.
- Confidentiality must be maintained throughout and after the research's conclusion. This includes the identity of the participants as well as the results of any testing or evaluation regarding individual performance.
- Detailed guidelines pertain to professional integrity regarding the integrity of data, its collection, and its distribution.

More questions? See #21, #23, and #24.

QUESTION #21

What Is Informed Consent, and What Does It Consist Of?

Informed consent is a process through which potential participants in a research study consent to a minimum set of standards that includes an understanding of what a research study is about, what role the participant plays, potential risks and benefits, and what the participant's rights are. The notion of informed consent was first used in the medical community during the late 1950s and has its beginnings in an 1891 Supreme Court declaration that individuals have a right to self-determination. It has been adopted by almost every organization whose professionals deal with research involving humans. Although informed consent cannot be obtained with animals, parallel procedures are used to ensure that animals are not treated cruelly.

The informed consent procedures are usually part of the Institutional Review Board's (IRB's) responsibility. The IRB reviews the informed consent form and procedures that a researcher plans to incorporate into his or her study (before the study begins, of course) and provides feedback when necessary. And, although the actual informed consent form differs from study to study and from institution to institution, they all include at least the following:

- The purpose of the study
- An option not to participate
- The procedures that will be carried out as part of the research and the amount of time that will be involved
- Possible risks
- Possible benefits
- Alternatives to participation
- Statements of confidentiality
- Additional information as the program progresses and complete disclosure of the results (of the entire group), if so desired by the participant.

All of the above need to be presented in a manner in which the language is fully understandable and not overly technical, and the potential participant has to be given adequate time to read and understand what is involved.

More questions? See #20, #24, and #25.

QUESTION #22

What Special Attention Should I Give to Ethical Concerns When Children or Special Populations Are Involved, and What Should the Parents or Legal Guardian Know?

Obtaining informed consent is an essential element of the research process, and every well-trained researcher realizes that it is his or her responsibility to ensure that it is obtained. The topic of informed consent with children, as with adults, has its origin in the medical community with the work of Dr. William G. Bartholomew and the American Academy of Pediatrics in 1985.

But, when dealing with special populations of potential participants, such as children or people who have limited capacity to communicate or understand, a special group of considerations comes into play. Almost all children do not have the legal right to agree to their own "treatment," so their parents or legal guardians have to make those decisions. In effect, informed consent for a child (or for anyone in the custody of an adult) is "consent by proxy." In other words, the process for requesting informed consent for a child looks very much like that used for an adult, and the parent or legal guardian is the one who is asked to consider the form's various sections.

When children are older than 6, some researchers use what is called an "assent" form or one where the research is explained in age-appropriate terms. This is not a legal document but an attempt to further involve the child and to inform him or her of the various aspects of the research. Here's a sample of what one form might look like for children ages 7 through 12 years.

Sample Child Participant Assent Form

Study Title: The Effects of Advertising on Toy Choice

Dr. Williams is doing a research study to find out whether the kinds of advertisements you see on TV have an impact on what kinds of toys you might want to get as a gift. If you decide that you want to be in this research study, this is what will happen to you:

1. Dr. Williams will show you some advertisements for a variety of toys.
2. You will then be shown these toys and allowed to play with them for 5 minutes.
3. Dr. Williams will then ask you which toy you would most like as a gift for your next birthday.

Sometimes, children who participate in this study might feel like this:

- They don't like any of the advertisements.
- They don't like any of the toys and feel bad because they think they should.
- They feel bad because they want the toys they played with.

If you feel any of these things, be sure to tell Dr. Williams or your mom or dad, and your mom or dad can tell Dr. Williams.

If you do not want to, you do not have to participate in this study. Nobody will be upset, and even if you do participate and want to stop at any time, that's OK.

Be sure to ask Dr. Williams or any of his helpers about anything that you don't understand, or let us know if you have any questions or concerns. Please let us know if you will participate by checking yes or no below.

___ Yes, I will be in this research study.

___ No, I don't want to do this.

____

Please write your name on the above line.

____

Date

More questions? See #20, #23, and #24.

QUESTION #23

What Are Some Examples of the Most Serious Ethical Lapses?

Unfortunately, there are many examples, and it would be very difficult to select the most serious, but here is a brief list compiled by David Resnick from the National Institute of Environmental Health Sciences that will give you some insight into how things can go very wrong. The violations and ignoring of participants' rights are blatant throughout.

You can find out more about these online.

- The infamous Tuskegee Syphilis Study (1932–1972) was sponsored by the U.S. Department of Health and studied the effects of untreated syphilis in 400 African American men. Researchers withheld treatment even when penicillin (a cure) became widely available, and researchers did not tell the subjects that they were in an experiment.
- German scientists conducted research on concentration camp prisoners from 1939 to 1945.
- From 1944 to the 1980s, the U.S. government sponsored secret research on the effects of radiation on human beings. Subjects were not told that they were participating in the experiments.
- James Watson and Francis Crick discovered the structure of DNA in 1953, for which they eventually shared the Nobel Prize in 1962. They secretly obtained key data from Rosalind Franklin without her permission.
- From 1956 to 1980, Saul Krugman, Joan Giles, and other researchers conducted hepatitis experiments on developmentally disabled children at The Willowbrook State School, where they intentionally infected children with the disease and observed its natural progression.
- The CIA administered LSD to unknowing participants in the early 1950s.
- In 1962, Stanley Milgram tested and proved that many people were willing to do things that they considered to be morally wrong when following authority.

More questions? See #20, #24, and #25.

QUESTION #24

What Is an Institutional Review Board or IRB, and How Does It Work?

In almost every institution where research is performed, be it a private or public university or even a corporation, an Institutional Review Board (or IRB) is convened to review the nature of the research.

An IRB is most often a diverse group of scientists from the institution in which the research is taking place, all of whom are involved in research activities. At a university, the board might consist of up to 10 faculty members, all of whom are involved in the research process, regardless of the discipline.

The primary task of the IRB is to review research proposals before the research begins, and to approve that the proposed methods to be followed do not in any way place the participants in danger. In addition, if an experiment involves deception, then they ensure that participants will be debriefed when the experiment or study is concluded. If children are involved as a special case of participants, then additional assurances have to be provided in order to conclude that this special group of participants is protected.

The process goes something like this.

The principal investigator (or P.I.), the person who will be in charge of the research should it be approved, completes an IRB application. This is not only required by the P.I.'s home institution, but if the research is to be funded by an outside agency (be it public, such as the federal government, or private, such as a private foundation), IRB approval is a critical aspect of the application.

Next, once the application is complete, and its details vary from institution to institution, the application is reviewed by the IRB committee. They then sign off on the application or make recommendations for changes.

If there are no concerns, the IRB approves of the research and that aspect of the process is finished. If the IRB has questions or doubts as to whether the participants are fully protected, the P.I. will be asked to review the application and resubmit, and in some cases that pose especially egregious threats to the safety of the participants, the application will be denied.

Special populations of potential participants, such as adults who are incapable of making decisions or minor children, require special consideration.

More questions? See #20, #22, and #23.

QUESTION #25

What Are the Important Elements of an IRB Application?

Every institution, including universities and corporations, has its own IRB forms, and many have different forms based on location (a hospital or a school, for example) and different types of participants (of age adults or children or special populations, for example).

But, IRB forms all tend to have many elements in common, and here's a brief listing of what you would need to take into account as you fill out such an application at your home university before you begin your research activity.

1. Location of research and specific site of research activity
2. Additional committees that may have to approve of the IRB application (such as a local public school as well as the university committee)
3. Information regarding the sources of funding for the research internal to the university, industry, etc.
4. A 250 (or so)-word abstract that explains the purpose and methods involved in the study
5. Background information to the study, including the rationale, hypothesis, and methodology, given in much greater detail than the briefer abstract described above
6. A very detailed description of the sample population
7. Special considerations given the population (such as incarcerated, needing special transportation, etc.)
8. How non-English instructions will be presented and, if necessary, whether a translator will be present
9. Participant identification information and the criteria for inclusion and exclusion, including the steps used to make contact with potential participants
10. Detailed description of methods

11. Specific demands or requirements of subjects (what they will be doing)
12. If necessary, how debriefing will take place
13. How risk will be minimized
14. Statement of any potential conflicts of interest

These 14 are minimal requirements and show up on almost every IRB form. Additional information is likely to be required.

More questions? See #20, #22, and #24.

PART 4

RESEARCH METHODS: KNOWING THE LANGUAGE, KNOWING THE IDEAS

QUESTION #26

Why Do All These Questions and Answers on Research Methods Have Any Relevance for Me?

Everywhere professionals look, there are controversies over which treatment, program, or method is best to teach one thing or the other, to help parents do a better job, to help nurses use their already limited time better, and on and on.

Such points for discussion are studied by researchers and eventually debated by decision makers every day. And, no doubt, you are confronted with numerous decisions about complex issues that will have a direct impact on who you work with—you hear all kinds of things from colleagues about why, for example, one method of teaching reading or increasing parent involvement is better than the other, but you also hear the opposite from other colleagues.

No matter what you hear, you're still faced with the responsibility of making a decision because you want to make one that "makes sense"—that works for you and your learners. And that's exactly where having an understanding of research methods fits in.

Research methods is the group of tools and techniques used to answer questions in a scientific manner. As you learn to understand the particular terminology of research methods and how to use these tools, you will become a better decision maker and a more effective administrator. For example, we're pretty sure that children whose parents are involved in their everyday lives do better in school and better in general. They get better grades, stay out of trouble, and are more industrious. Are those statements based on speculation? Nope.

The results of extensive amounts of research studies have shown this to be the case. Now, the next time a parent asks you why it matters if he shows up at his son's teacher's conference (and he really is looking for an answer), you have it, and you have it because many educational researchers approached this question systematically and answered it.

So, here's what all the questions and answers in *100 Questions About . . .* can do for you . . .

- Help you better understand the process of asking and answering a question systematically.
- Provide you with the tools to understand and analyze the contradictory information thrown at you by parents, consultants, and everyone else who wants you to do as they say—not what is necessarily right.
- Be a better consumer of the kind of information that you really need to be the best professional you can.
- And finally, just be really smart—and learn how the increasingly large amounts of information out there can be better understood and acted upon.

More questions? See #3, #6, and #10.

QUESTION #27

I Have So Many Ideas I Want to Study. How Can I Decide Which One Is Best?

Many different considerations should enter into your decision as to which research question you will try to answer. Here's a summary of the 10 most important.

1. Try to select a problem that is within your discipline. This will allow you to tap your adviser's knowledge as well as complement your coursework or your work interests.
2. Don't fall in love with your idea. You certainly want to be passionate about what you are exploring, but as in all cases of love at first sight, you don't want your judgment to be based only on emotions.
3. Go beyond the first idea you come up with. That first idea is always going to be a kind of "Gee Whiz" moment based solely on how cool something would be to do. But give yourself room to explore variants of that first idea or even move on to others.
4. Do something significant and something that is based on previous literature and contributes to a further understanding of your topic.
5. Don't undertake a 4-year, 2,000-participant study of 10 variables. Be realistic, and select a topic and ask a question that you can answer in a reasonable amount of time. Be ambitious, but be reasonable as well.
6. Look at what's been done in the past and build on that. Everyone's work, in one way or another, stands on the shoulders of those who came before.
7. Work closely with your adviser. He or she knows the field far better than you, and spending 30 minutes every 2 weeks, after you have done your reading of articles in your area, presents a great opportunity to expand your horizons.

8. There's nothing like going to the (real) library. Online access is just fine, but you just can't see which article is "next to" the one you're looking at online.

9. Bounce ideas off your colleagues. Even if they are relatively new to this venture, they may have some insights that are very useful.

10. Finally, look upon this whole effort as an exploratory adventure. You're doing this to grow professionally and intellectually, so allow plenty of time to explore and keep good notes.

More questions? See #14, #16, and #26.

QUESTION #28

In Beginning My Research Work, Can I Focus Just on One Tiny, Little, Narrow Topic or Reach for the Stars and Be Broad and General? And, I Know the Library Is a Terrific Place to Start My Research Work, but Do I Have to Visit the Bricks-and-Mortar Buildings on Campus or Can I Just Work Remotely?

Well, especially for a relative beginner in any scientific endeavor, you'd be better off reaching a bit for the stars while still knowing the general area in which you want to work. If you are too general, you will never be able to find your way because you could be reading for years until something comes up that fits your interests. If you are too narrow, all you will see are the trees and not the forest—you'll miss the big picture.

But much more important, when you find a middle ground where you read about what interests you (with predefining exactly what you will study), you have the luxury (and it is a luxury) of thinking about important and "big" topics and ones that have great potential for further study.

As far as online research versus visits to campus or downtown to the library, actually, you should seek out both when you do your preliminary literature review as well as accumulate any information that you need later on. Nothing can beat online searches when you know *exactly* what you are looking for. If you need references on the relationship between learning style and achievement, you can probably very well find what you want and need online.

However, there's nothing like a trip to the library (and the stacks); ending up in the general area of your interests; and then looking through journals, anthologies (collections of chapters or essays on a related topic), and, by chance, finding the book next to the one that you are looking at, or a journal article you never heard about that piques your interest further and even might have a key point that fits nicely into your argument and your work. Try both—and you never know whom you'll meet at the library as well.

More questions? See #10, #12, and #18.

QUESTION #29

What Is a Null Hypothesis, and Why Is It Important?

A null hypothesis is a statement of equality. For example, the statement that the averages of two groups are equal, that two variables are unrelated to one another, or that the difference between two groups is equal to zero are all examples that would reflect a null hypothesis.

Null hypotheses have two purposes.

First, they are a starting point. Given no other information about the relationship between two variables (as to whether they are related or not, for example), researchers must assume at the beginning of their work that the variables are not related. It is the most neutral and unbiased position from which to begin a research endeavor.

Second, a null hypothesis is a benchmark against which findings (the results of the actual test of the research hypothesis) can be compared. It's the standard against which the outcomes of the actual experiment can be compared and a decision can be made as to whether these results are substantially different from the null.

Let's see how both of these might work in an example.

Suppose that a researcher is looking at the relationship between quality of work habits and level of productivity, and he or she uses a reliable and valid measure of both of these variables. One appropriate null hypothesis would be that there is no relationship between work habits and level of productivity.

The researcher's hypothesis (and there can be more than one) is that these two variables are related such that if the quality of work habits increases, so does the level of productivity.

The null acts as a starting point because, knowing nothing else about the relationship between these two variables, the researcher has to assume that the quality of work habits and level of productivity are unrelated and that change in one does not affect the other. And, this null also has to act as a benchmark because it is the reference point against which the researcher will compare the actual findings based on data collected to see if, indeed, the difference is greater than one would expect by chance alone.

More questions? See #8, #9, and #30.

QUESTION #30

What Is a Research Hypothesis, and What Are the Different Types?

You already know that a null hypothesis is a statement of equality and that it applies to populations that cannot be tested directly.

A research hypothesis is the other side of the "hypothesis coin." A research statement is a statement of inequality, and there are two basic kinds.

First, there is a nondirectional research hypothesis, which states that two sample statistics (such as two averages) are different from one another. In this case, the null hypothesis would be that there is no difference. An example of a nondirectional research hypothesis would be that the effects of cognitive therapy on self-injurious behavior are different from the effects of psychotherapy. In research speak, you might see it expressed as . . .

$$H_1 : \bar{X}_{CT} \neq \bar{X}_P$$

where H_1 stands for the first research hypothesis (of which there can be more than one), *CT* stands for cognitive therapy, and *P* stands for psychotherapy.

Second, there is a directional research hypothesis, which states that two sample statistics (such as two averages) are different from one another, but there is a direction (more ">" or less "<" than) to the difference. An example of a directional research hypothesis would be that the average age of retirement for men is greater than the average age for retirement for women. In research speak, you might see it expressed as . . .

$$H_1 : \bar{X}_{Men} > \bar{X}_{Women}$$

What do research hypotheses do best? They guide our research in that they reflect the initial questions that are asked as well as the goal of the experiment of which they are a part. In any one research study, there can be multiple experiments of all different levels of complexity, and the more variables that are involved, it is likely that there is more than one research hypothesis and that the research hypotheses become more complex.

More questions? See #3, #9, and #29.

QUESTION #31

What Is Similar, and What Is Different, About a Null and a Research Hypothesis?

To begin with, here's what's similar.

Both the null and the research hypotheses are statements about relationships between variables. In the case of a null hypothesis, it is statement of equality (for example, the average score for Group 1 will equal the average score for Group 2). In the case of a research hypothesis, it is a statement of inequality (for example, the average score for Group 1 will exceed the average score for Group 2).

And, that's where the similarity ends. Here's how they are different . . .

1. The null hypothesis is the default hypothesis in any research study. The research hypothesis is often referred to as the alternative hypothesis.

2. Because it refers to the population, a null hypothesis is never tested directly, whereas a research hypothesis is. Whether a null hypothesis is "true" or "false" is implied from the results of a test of the research hypothesis. One cannot directly observe that a null hypothesis is true or false.

3. A null hypothesis is expressed using population parameters and Roman symbols (such as σ [sigma] and α [alpha]), whereas a research hypothesis is expressed using sample statistics (using Greek symbols such as *s* and *a*).

4. A null hypothesis always refers to a population, whereas a research hypothesis always refers to a sample.

5. A research hypothesis is tested and has a level of error or a significance level associated with it.

6. A null hypothesis acts as a starting point in a research study, where, given no other information, it is assumed that there is no difference between variables.

7. The null hypothesis acts as a benchmark against which results are compared. The results that are compared are derived from a test of the research hypothesis.

8. A research hypothesis produces that which is compared to the expected results of the null hypothesis (which is no difference).

More questions? See #3, #29, and #30.

QUESTION #32

How Can I Create a Good Research Hypothesis?

There are several criteria that make a research hypothesis a good one, and following them is a very good start.

1. A good hypothesis is stated in declarative form and not as a question. "Are swimmers stronger than runners?" is not declarative, but "Swimmers are stronger than runners" is.
2. A good hypothesis posits an expected relationship between variables and clearly states a relationship between variables. For example, the research hypothesis "Children who participate in three hours of lap reading with parents per week will score higher on a test of reading comprehension than children who do not" states a clear relationship between hours of reading and test score.
3. Hypotheses reflect the theory or literature on which they are based. A good hypothesis has a substantive link to existing literature and theory. In the above example, let's assume there is literature indicating that reading to children is one way to increase their comprehension. The hypothesis is a test of that idea.
4. A hypothesis should be brief and to the point. You want the research hypothesis to describe the relationship between variables and to be as direct and explicit as possible.
5. Good hypotheses are testable hypotheses. This means that one can actually carry out the intent of the question reflected by the hypothesis. For example, number of hours of parental reading and outcome scores as measured by a test of comprehension are all objective and can be incorporated reliably.
6. Finally, a good research hypothesis combines all of the above to be understandable and easy to envision how it fits into the larger world of the research question. After reading such a hypothesis, the reader should have a good grasp of which direction the research is taking and what some of the implications for its testing might be.

More questions? See #9, #29, and #30.

QUESTION #33

What Is the "Gold Standard" of Research Methods?

There are many different measures by which you could evaluate the effectiveness or potential value of any research method, especially in disciplines where there is a premium placed on experimental outcomes. The scientific method is the gold standard and is the standard against which methods for studying new questions can be compared.

Basically, the scientific method requires that observable or empirical data be collected through experimentation and that these data reflect a hypothesis or some educated guess. In more detail, the scientific method consists of six steps.

- Step 1 is asking a research question that the researcher would like to have answered once the study is completed.
- Step 2 is completing background research in which previous work on this same topic is reviewed and a comprehensive perspective is offered on how the new research is an extension of the previous work.
- Step 3 is creating a research hypothesis that defines the variables to be tested.
- Step 4 is testing the hypothesis, including a method that clearly addresses the hypothesis and an objective (as possible) definition *of what variables will be tested and how.*
- Step 5 is analyzing the data and outlining the conclusions that might be drawn from the analysis.
- Step 6 is sharing the results with the entire community of researchers that may be interested, including a statement of future avenues for further research.

Perhaps what is most interesting about the scientific method is that it is not necessarily tied to any one type of research model, such as the different quantitative or qualitative models that now are used. It is all about the integrity of observation and testing as a way of revealing information not previously known, regardless of the general philosophy behind the method.

More questions? See #3, #4, and #31.

QUESTION #34

Can You Help Me Understand Which Method Best Fits Which Type of Question Being Asked?

As you already know, different research models or types of research are often used to answer different types of questions. For example, correlational models are used to determine if there is a relationship between variables, such as between a test of conceptual thinking and number of hours practicing those kinds of items. Or, an experimental model that looks at the cause-and-effect relationship between sets of variables might examine the effectiveness of a particular type of intervention program to reduce alcohol consumption.

The quick summary is something like this . . .

If you are interested in . . .	Consider this method as the primary method . . .
Describing the current nature of things	Descriptive
Describing how past occurrences relate to present events	Historical
Understanding the relationship between variables	Correlational
The unique elements and social context of individuals, institutions, or phenomena	Qualitative
The cause-and-effect relationship between variables	Experimental

But do keep the following in mind as you let your research question guide you toward a hypothesis and then toward a method.

1. The research questions should always drive the method selected and not the other way around.
2. These models are to be considered a starting point and a framework within which you will design your particular study, specifying a set of variables and identifying the possible relationship between them.
3. The second column in the above table contains the word *primary* because there are always questions that require the use of more than one method to reach a satisfactory answer.
4. No method stands alone, and there are almost always elements of more than one method in any one research study.

More questions? See #4, #15, and #38.

QUESTION #35

What Are the Different Types of Variables, and What Are They Used For?

A variable can be defined as anything that can take on more than one value.

For example, the number of postsurgery infections in a hospital and the number of workers receiving unemployment in Milwaukee are variables, as is the number of cars produced by Toyota in its newest manufacturing plant. What is not a variable is a constant, something that cannot take on more than one value. For example, your street address does not change—it's not variable, nor is it a variable.

Variables are what scientists study, and there are many different kinds. Later in *100 Questions . . .* (the next question, in fact), we'll go into a bit more detail, but for now, here are some basics about different types.

Independent variables are those that are manipulated in experimental or quasi-experimental studies, or are defined as having several levels, such as different levels of grade. It's what researchers define to test whether there is a difference among the levels.

A difference in what? Whatever it is they are interested in observing an effect upon. This outcome variable is also known as the dependent variable. So, if a researcher was interested in seeing whether there is a difference in grade level in the number of friends that 6th, 9th, and 12th graders have, then the number of friends would be the dependent variable and the three different values of grade would be the independent variable.

Another important type of variable is a moderator variable. A moderator variable is one that is related to other variables, and it affects the strength of the relationship between those variables. For example, gender (male or female) may be a moderator variable if it relates to both muscle strength and age of onset of puberty. Muscle strength is related to age of onset of puberty, but because it is related to both, gender becomes a moderator variable.

More questions? See #9, #36, and #38.

QUESTION #36

What Is An Independent Variable, and How Is It Used in the Research Process?

As you already know, there are many different types of variables, and they all play an important role in a research study.

An independent variable is one that has several different levels reflecting the variety of treatments that the researcher wishes to test the effects of on an outcome or dependent variable. For example, a researcher wants to see if there is a difference in sales as a function of three different modes of print advertising: color, black and white, and a combination of both. Here, there is one independent variable—mode of advertising—and there are three levels of the one independent variable. When the research is completed, the researcher should be able to conclude whether there is a difference between these three levels and, if there is, which one resulted in the most sales. Visually, the design would look something like this . . .

One-Factor Design

Mode of Advertising		
Color	**B&W**	**Color and B&W**
Number of Sales	*Number of Sales*	*Number of Sales*

The above example is one where there was only one independent variable (called a one-factor design), but research studies very often have more than one independent variable. For example, as shown below, the same researcher might be interested in looking at the effect of different modes of advertising on sales, but is also interested in a second variable or factor—level of literacy. This would be a two-factor or two-dimensional design. One factor (mode of presentation) has three levels, and the second factor (level of literacy—high, medium, or low) has three levels as well. This is commonly known as a two-dimensional design (with two factors) with three conditions within each factor of the 3×3 design.

In all research studies, when independent variables are involved, they are there to test the effects of their different levels on some outcome variable and are often described visually as follows.

Two-Factor Design

		Mode of Advertising		
		Color	**B&W**	**Color and B&W**
Level of Literacy	High	*Number of Sales*	*Number of Sales*	*Number of Sales*
	Medium	*Number of Sales*	*Number of Sales*	*Number of Sales*
	Low	*Number of Sales*	*Number of Sales*	*Number of Sales*

More questions? See #35, #37, and #38.

QUESTION #37

What Is a Dependent Variable, and What Does the Researcher Need to Be Careful About When Selecting and Using Dependent Variables?

A dependent variable is used to determine if one (or more) independent variables had an effect. A dependent variable can also be a measure of an outcome that stands alone, such as the results of a survey in which there are no manipulation or independent variables present.

Several things are important to keep in mind when considering the selection and use of a dependent variable.

First, it is ideal when the dependent variable is related to, and sensitive to changes in, the independent variable. This way, the researcher is as assured as possible that the effect, if there is one, of the independent variable will be apparent. If one were to look at the impact of an exercise program on teenage obesity, weight gain would seem to be a very relevant and appropriate outcome. You can well imagine looking at the impact of an exercise program on teenage obesity and using a dependent variable such as college aspirations. It's unlikely that weight gain and college aspirations are closely (and meaningfully) related.

Second, if more than one dependent variable is used in a research study, then they should be as unrelated to one another as possible. Weight loss would be a very good dependent variable for the above-mentioned study, but using another, similar dependent variable, such as calorie intake, could be redundant and unproductive.

Finally, whatever dependent variable is used, it must be a reliable and valid measure of what is of interest. Using a dependent variable that is unreliable would not accurately reflect whether the treatment would be effective. Why? If the variable is unreliable (and therefore invalid), the researcher does not know whether change (or lack of change) in the dependent variable is a function of the poorly constructed dependent variable or actually an effect of the treatment.

More questions? See #35, #38, and #39.

QUESTION #38

What Is the Relationship Between Independent and Dependent Variables?

It's a very interesting one and most simply expressed and understood using the following equation:

$$\text{dependent variable} = \text{function of } (\text{independent variable}_1, \text{independent variable}_2 \ldots)$$

or

$$dv = f(iv_1, iv_2, \text{etc.})$$

What this formula represents is that the value of the dependent variable is a function of changes in one or more independent variables. In this equation, there are two independent variables, but in theory, there can be three, four, or many more, but we'll get to that in a moment.

For now, you should understand that the relationship is a special one in several ways.

First, this formula expresses the bare bones of the experimental method, in which one or more independent variables are tested to see if they have an effect on the dependent variable. For example, a researcher might be interested in testing whether a change in employment status has an impact on self-esteem.

Second, each of the independent variables should be independent of one another. This is to ensure that they each contribute unique information to your understanding of changes in the dependent variable. For example, if a researcher were interested in studying employment status, he or she would probably not want to include income because most unemployed people have no current income stream and most employed people do. Employment status and income stream provide a similar type of information, so there is a great deal of overlap.

Third, independent variables can often interact with one another. For example, employment status and age may not be related individually to level of self-esteem, but working together (such as younger unemployed participants versus older employed participants), there very well might be an effect on the dependent variable.

Finally, one might think that the use of many independent variables is the answer, because more is better in that they explain more. The problem is that, eventually, independent variables tend to "repeat" each other and overlap (more of the second reason above), so it becomes expensive to collect such redundant data and a marginal waste of time and resources.

In sum, a researcher wants dependent variables that are sensitive to changes in the independent variable(s), but independent variables that are unrelated to one another.

More questions? See #35, #39, and #84.

QUESTION #39

In an Experiment, How Does the Notion of a Control and an Experimental Group Fit Into the Scientific Method?

Experimental research is often the most frequent kind of research model used to test for cause-and-effect relationships in the social and behavioral sciences. The form of this type of research often takes the following steps.

1. A population of potential participants is identified.
2. A sample or samples are selected from the population in such a way as to ensure as much as possible that the samples represent the population.
3. Each of the participants in one of the samples is assigned to an experimental group that receives one or one of several treatments.
4. Each of the participants in the other sample is assigned to the control group, which does not receive the treatment.
5. Once the treatments have concluded, each participant in each group receives some type of assessment in an effort to see whether the exposure to an experimental condition resulted in a difference in scores between the experiment and the control groups.

The logic behind this type of strategy is that both the experimental and control groups having been selected from the population share all of the same characteristics, biases, predispositions, and so on. If the population consists of 10% unemployed members and 47% males, then each of the samples should reflect that as well (if the sampling is done correctly).

So, if one group is exposed to a condition (called the treatment) and the other is not (and is the control group), then any difference between the groups should be attributable to the effect of the treatment.

This is a pretty basic explanation and somewhat simple as well, but the logic holds tight for one or several experimental groups. If done correctly, a test of differences between all groups should reveal whether or not the treatment is related to the outcome.

More questions? See #3, #33, and #84.

SAMPLING IDEAS AND ISSUES

QUESTION #40

What Is the Difference Between a Sample and a Population, and Why Are Samples Important?

Samples are selected from populations.

A population is the total of all the individuals who have certain characteristics and are of interest to a researcher. Community college students, race car drivers, teachers, college-level athletes, and disabled war veterans can all be considered populations. Because sampling is not a perfect part of science, there are often differences between the values of a sample and the values of a population. This is called sampling error, and it is the researcher's duty to minimize this type of error.

A sample is a subset of the population. In the above example, only community college students in three schools in New Hampshire would constitute an appropriate sample, as would only veterans who incurred a specific type of injury during the Vietnam War.

The reason why samples are important is that within many models of scientific research, it is impossible (from both a strategic and a resource perspective) to study *all* the members of a population for a research project. It just costs too much and takes too much time. Instead, a selected few participants (who make up the sample) are chosen to ensure that the sample is representative of the population. And, if this is the case, then the results from the sample can be inferred to the population, which is exactly the purpose of inferential statistics—using information on a smaller group of participants to infer to the group of all participants.

There are many types of samples, including a random sample, a stratified sample, and a convenience sample (more about those later), but they all have the goal of accurately creating a smaller subset from the larger set of general participants such that the smaller subset is representative of the larger set.

More questions? See #39, #41, and #42.

QUESTION #41

What Is the Purpose of Sampling, and What Might Go Wrong During the Process?

Sampling has one primary purpose: to select a group of participants that is representative of the general population of all possible participants. Samples are always smaller than populations, and we strive for the sample being selected to be representative of the population.

Representation is the basis of inferential statistics, where the results from a sample are generalizable or applicable to the population from which the sample is selected.

For example, if a researcher were interested in studying the effects of an exercise intervention program on 60- to 70-year-old sedentary adults, she would select a sample of a particular size from the much larger population of all possible adults between the specified ages who work in sedentary jobs. The groups that are formed from the sample would then be exposed (or not, in the case of the control group) to the treatment. Once the experiment is concluded, the researcher would then be able to take the results of the study based on the sample and generalize them to the population. The more precise the sampling (that is, the more accurate), the more applicable are the results to the population. In fact, the results of any study where samples are used are only as generalizable as the accuracy with which those samples are drawn.

What might go wrong? Quite a few things, but among the most serious are the following two.

First, the sample might be selected incorrectly and not be representative of the population. For example, rather than selecting participants in a way that ensures an equal chance for all, the first 30 participants might be selected because the researcher knows these individuals.

Second, the sample can be far too small, increasing the likelihood that the study's participants are not representative of the population.

Both of these errors can lead to imprecision in generalizing back to the population (the goal of almost all inferential statistics).

More questions? See #28, #39, and #42.

QUESTION #42

What Is Sampling Error, and Why Is It Important?

Sampling error is the difference between the value of a sample statistic and the value of a population parameter. In other words, if you were to calculate the average SAT score for a group of students from one high school in Cleveland (the sample), there would be a difference between that value and the average SAT score for all students throughout the United States (the population). That difference is a measure of sampling error and reflects how the sampling process is never perfect. (It can also reflect a genuine error in sampling correctly.)

Sampling error, in theory, is an average of all the differences between all the possible samples and the population measure. The lower the value of the sampling error, the more precise the sampling process is. If the value were equal to zero, sampling would be perfect and the sample would match the population perfectly on the variable of interest without any concern for differences.

Sampling error is based on the standard deviation of the samples measured (how much variability there is among them) and the size of the sample. The smaller the variability of the samples (the more alike they are), the smaller the sampling error. This is because the less variability there is among the many samples, the more the sample values, as a group, reflect the population values.

And, the larger the size of the individual samples, the smaller the sampling error as well. And, this is because as the size of a sample gets larger, it more closely reflects the qualities of the population. So why not measure the entire population? Too expensive and not necessary.

Samples should always be large enough to reflect the population (if the sampling is done correctly), but not too large that unnecessary time and resources are wasted.

More questions? See #41, #47, and #48.

QUESTION #43

What Are Some of the Different Types of Sampling?

Sampling is the process through which a group of participants is selected such that this smaller group or subset is representative of the entire population of all participants. It's important because we can conduct research using a smaller number of participants, and because if the sampling is accurate, those results can be confidently inferred to the population of participants.

There are many different types of sampling, and here's a summary of when to use the most important . . .

Type of Sampling	When It Should Be Used
Simple Random Sampling	When the population is homogeneous (very similar) and when the research question at hand does not require any attention to special characteristics of the population.
Stratified Random Sampling	When the population contains potential participants who have characteristics that are related to the variable under study. For example, when studying the development of early verbal skills in young males and females, care must be given to the selection of such samples because the variable of interest (onset of verbal skills) is related to gender (girls usually acquire such skills significantly earlier than boys).
Cluster Sampling	When a cluster or grouping of similar units, and not individual participants, is of interest, such as a group of nursing homes that represents all the nursing homes in the state.
Convenience Sampling	When a group of participants from whom a sample will be drawn is easily accessible and easy to select, such as an intact business, small manufacturing plant, or political party meeting.
Quota Sampling	When a specific size sample, regardless of other characteristics, is selected, such as the first 20 people to board the commuter bus in the morning.

More questions? See #40, #41, and #42.

QUESTION #44

What Is Random Sampling, and Why Is It So Useful?

There are many different ways to select a sample, and one of the best ways to ensure that the selection of participants is truly representative is by selecting a random sample.

Two conditions need to be met to ensure that a sample is random.

The first condition is that each of the participants has to have an equal chance of being selected as part of the sample. If the population consists of 1,000 women who are participating in a triathlon, then each of the 1,000 women has to have an equal chance of being selected. This means that the first 50 women to register for a race would not be a random sample because it means that the other 950 registrants never had a chance of being selected.

The second condition is that each of the participants has an independent chance of being selected. This means that if a researcher selects every 20th participant of the 1,000 registrants (which would total 50), it would not be a random sample because women numbered 1 through 19 and 21 through 49 and so on have no chance of being selected.

So, each participant in the population needs to have an equal and independent chance of being selected.

But why is random so much better than other sampling methods and, given no other constraints, the most preferred? Because a random sample is most likely to distribute any potential biasing characteristics across all the groups being formed through the sampling process. For example, if 65% of the women are experienced triathletes, we would expect that experience to be distributed equally across all groups (such as experimental and control) that are formed using random sampling.

More questions? See #40, #42, and #48.

QUESTION #45

How Does Stratified Random Sampling Work, and When Should I Use It?

Stratified random sampling is a special type of random sampling. All the criteria of random sampling still apply (independence and equal likelihood of being chosen), but there's another dimension as well. When the population from which a sample is being drawn consists of different strata that (and this is the most important part) have some relationship to the variables being studied, then one has to stratify the sample.

For example, let's say a researcher is interested in the political aspirations of a group of young adults ages 25–34. It is well known in the literature that young adults of different genders tend to have different views of politics, as do young adults of different economic classes, and it is the same as well with another variable, level of education. And, let's assume that the important outcome variable is attitude toward government.

Because these three factors (gender, economic class, and level of education) are related to one's attitude toward big government (as reported by other related studies), you would want the proportion in the partial sample of male, low education, and low economic class to match that of the proportion of young adults in the generally population from which the sample was being collected.

The different variables are like the layers of rock in stratified rock, and by taking a core sample (the sample) of that rock, one should get similar proportions as those present in the entire field of rock (the population). The reason why stratified random sampling is so effective is that it takes into account the effectiveness of random selection and addresses the circumstance in which participants with different, and related, characteristics are addressed as well.

More questions? See #41, #42, and #43.

QUESTION #46

How Can I Be Sure That the Sample of Participants, Which Is Part of a Study, Accurately Represents a Larger Group of People for Whom Those Results Would Be Important?

This is exactly the challenge that any researcher faces when first designing an experiment. For the results of research to be maximally generalizable to the larger population, the goal is to select a sample that is as similar as possible to the population. How can you do that? Here are some things to be sure of . . .

1. You need to have your research question, hypothesis, and review of literature complete and be able to identify whom you want to participate in your research. If you are looking at aspirations of teenagers to find a job after high school, for example, then you have to focus on exactly that population.
2. You need to make sure you know the characteristics of the population from which you will be selecting. So, although you may be interested in teenagers who will be looking for a job shortly, you want only those who actually aspire to a job as opposed to those who do not show any interest.
3. You need to make sure that there are few intervening variables that may clutter up the integrity of your sample. In other words, you want it to be as homogeneous as possible—so the members of the population are as alike as possible.
4. Finally, if possible, you want any outliers (those exceptionally weak or strong candidates, such as the one who aspires but will never be hired or the one who aspires and already has a job) eliminated from the pool of participants from which you will making your selection. Such outliers might bias the sample and threaten how generalizable the results are to a population that does not contain such outliers.

More questions? See #41, #43, and #48.

QUESTION #47

I've Heard Quite a Bit About the Importance of Sample Size. What's That All About?

The question always arises as to how big a sample needs to be to generalize successfully to the population, and the answer depends on many different factors. What might be some of the factors that determine how big a group you should use?

1. Be practical. If you can locate participants so that you have somewhere around 30 in each group, you're doing fine and you can limit your sample size to that.
2. Look to other studies that have asked questions similar to yours. The published literature is your best reference for what's right and what's not right to do. Given the topic you are researching and the characteristics of your sample, you can use previous work as a guideline to how large a sample is appropriate.
3. The larger the differences you expect between groups, the relatively smaller sample size you need for each group. This is because larger differences are more obvious than smaller ones, right? For example, if you are interested in seeing whether the height of 1st graders differs from the height of 6th graders, you surely don't need 30 children in each group. Rather, you can use three or four because it is very unlikely that in typically developing children, there are any 1st graders who are taller than the shortest of 6th graders.
4. Conversely, as the suspected difference between groups gets smaller, you need larger samples to show if any difference really exists. For example, if you suspect that a language immersion program will increase language skills but not by a huge amount, you will need a larger sample (closer to the population size, right?) to show that difference.

5. Remember that this is the real world. If you are dealing with special groups, such as children with disabilities, or adults older than 90, or fathers who are stay-at-home dads (still a small percentage of parents with full reasonability), you will be hard pressed to find lots of participants. Do the best you can under these circumstances and realize that the generalizability of such results may be somewhat limited.

More questions? See #41, #42, and #44.

QUESTION #48

How Big of a Sample Is Big Enough?

This is an important question because most beginning researchers think that bigger is always better and that the larger the sample, the better the final results of the research. That's not the case.

It is important to remember that the larger the sample, the more representative it probably is of the population and the smaller the sampling error, but there are other important considerations to keep in mind.

To begin with, larger samples use more resources such as time and money. It takes more time to assess more participants, travel to the research site, and analyze the data, and countless other expenses are incurred. Why incur those expenses when one does not have to?

Second, there are definitive factors that affect the impact of a large or small sample on the final results of one's research.

One is the amount of variability in the sample, or samples, being investigated. For example, if there are two groups of participants, one that receives a treatment and one that does not, and if the variability within each of these groups is quite small (and they are homogeneous or very much alike), a smaller rather than a larger sample might do just fine. Why? They are so alike to begin with that few "real" differences would be needed to be considered significant or meaningful.

Another factor is the previous research that has been done and how large the samples were in those studies. This information should be an important source of guidance and a good starting point.

Finally, results of previous research become even more important when one considers how large samples were and whether differences were found. If a study looked at differences between groups and the sample size was 50 in each group, and if, all other things being equal, a similar study is being planned, then 50 might be the perfect number with which to start.

More questions? See #41, #46, and #47.

QUESTION #49

How Important Is Big?

Does a small sample automatically reduce the value of the results of a study? Well, as we have said before, that depends upon the question you are asking. But, in general, the answer is no. If anything, a small sample may limit the degree of generalizability, but that just calls for the study to be repeated with a larger sample or perhaps one with different characteristics. More information means being able to make better decisions.

And, will a large sample make the results more useful? Nope. A large sample might just cost more to collect the data but not yield much more valuable information. Not too big and not too small—you want just the right size.

What you want is a study size that is adequate. One that allows you to test the hypothesis you propose and with which you will get an accurate and fair answer. So, don't necessarily look for big—look for accurate, reasonable, cost-efficient, and above all, doable. Big only counts on Thanksgiving.

In general, you want the size of a sample to be large enough to truly reflect the characteristics of the population, but not so large that you waste valuable resources (such as time and money) testing participants beyond what you need. And, in many cases, this magic number seems to be about 30 (per group), which is where the mathematics of generalizability (far beyond our interest here) kick in.

More questions? See #47, #48, and #92.

DESCRIBING DATA USING DESCRIPTIVE TECHNIQUES

QUESTION #50

What Are Descriptive Statistics, and How Are They Used?

Descriptive statistics describe outcomes, and there are two primary types of measures that are used: measures of central tendency and measures of variability.

Measures of central tendency provide the one best value (and it can be numerical or non-numerical) that represents a set of data scores. For example, the average speed to run 1 mile for a group of adolescents may be 6:47, or the one data point that represents a group of urban residents' position on a particular ballot vote might be "no." In this general category of descriptive statistics, you find descriptive statistics such as the mean, median, and mode. In almost any research report that involves the collection of numerical data, you can find one or more of these measures being reported.

Measures of variability provide a value that reflects the average amount of difference between each score from one another and is used to get some idea about how much spread or dispersion there is in a group of scores. For example, in two groups of adults (say, males and females), the average attitude toward receiving phone calls at home during the dinner hour (with the average being a measure of central tendency) might be identical, but females may be very similar to one another in their attitude and males quite different. In this example, males and females differ in the amount of variability.

What knowing these descriptive statistics does is it allows us to describe what a group of data points looks like and to make some initial and important comparisons between groups. Surely, most researchers go beyond just a description of outcomes, but this first step is an essential part in summarizing the outcomes of what's being investigated.

More questions? See #51, #53, and #55.

QUESTION #51

What Are Measures of Central Tendency, and How Are They Computed?

Any measure of central tendency provides the one most important data point for describing a set of data points. If you had no other information and wanted to know what the one best data point would be to describe spelling test performance for a class of 10th graders, you would use the mean.

There are three measures of central tendency that all researchers use.

The first is the mean, which is the arithmetic average of a set of scores. It is computed by adding up each score and then dividing the sum by the number of scores. The mean for the following set of scores

2, 2, 3, 5, 6, 6, 11

is 7. The mean is usually represented by a capital letter *X* with a bar over it ($\bar{X}$) or the capital letter *M*.

The second important measure of central tendency is the median, which is the score that marks the middle of the set of scores, or the point at which 50% of the scores fall above and 50% fall below. It is computed by simply finding the middle-most score, which, in the example, is 5. The median is used as a measure of central tendency when a set of scores contains extreme scores that would unfairly weight the mean such that it would not be the best reflection of one measure of central tendency.

Finally, the mode is the value that occurs most frequently in a set of scores. For example, if one were to describe a group of 100 workshop participants, one could say that 56 of them are females and 46 are male, and female would be the mode.

More questions? See #50, #52, and #57.

QUESTION #52

How Do I Decide Whether to Use the Mean, Mode, or Median as a Measure of Central Tendency?

Which measure of central tendency you use depends on the level of measurement of the variable or outcome you are analyzing. You remember that there are four levels of measurement: nominal, ordinal, interval, and ratio. Any outcome can be assigned to one of those levels and as the level of measurement goes from nominal to ratio, the degree of precision in measurement increases.

With that in mind, the following chart will give you some idea as to what measure of central tendency best fits what level of measurement along with an example of each.

Level of Measurement	Appropriate Measure of Central Tendency	Variable or Outcome	Example
Nominal	Mode	Political party affiliation as measured by preference in a survey where there are 20 Republicans, 15 Democrats, and 12 Independents.	The mode is Republican.
Ordinal	Median	Income level measured in dollars for a wide range of employees with some extreme scores.	Average or median income is $64,765.

Level of Measurement	Appropriate Measure of Central Tendency	Variable or Outcome	Example
Interval	Mean	Number of words remembered after an intervention program involving traumatic head injury.	Mean number of words remembered is 13.5.
Ratio	Mean	Number of green houses built each year for 5 years following a tornado.	Mean number of houses built each year are 6, 4, 7, 5, and 8.

You should keep in mind that, if possible, a higher level of measurement is preferable to a lower level of measurement because the higher level provides more precision and more information about the outcome. For example, knowing that female children outperform male children on certain measures of verbal skills is not as informative as knowing how wide this difference may be.

More questions? See #51, #54, and #57.

QUESTION #53

What Are the Most Often Used Measures of Variability, and How Are They Computed?

There are three often used measures of variability, all of which reflect the amount of spread or dispersion in a set of data.

The first is the range, which is simply the lowest score in a set of scores subtracted from the highest score in the same set. It is the most general and gross measure of variability that is used and is quick and simple to compute. It is not very accurate, because sets of scores with the same high and low score can have different levels of variability, such as the set 3,4,4,4,4,4,4,4,5 and 3,3,3,3,4,5,5,5,5—same range (and same average), yet different amounts of variability. The range is usually represented by the letter R.

The second measure is the standard deviation, the most often reported measure of variability. The standard deviation (*s* or *sd* for samples and s for populations) is the average deviation of each score in the data set from the average of the entire data. It is computed by subtracting each individual score from the mean, squaring that deviation, computing an average of the deviations and then taking the square root of that value (to return it to the original units). Standard deviations are reported along with means as two essential descriptive statistics.

The final measure of variability is the variance, which is the standard deviation squared or σ^2 for samples and σ^2 for populations. The variance as a concept is important in the understanding of certain inferential statistical tests such as analysis of variance.

When to use which? The range is an overall but imprecise estimate, whereas the standard deviation provides the most often used metric to complement the mean in understanding the basic appearance of a set of data.

More questions? See #51, #54, and #55.

QUESTION #54

How Do I Use the Mean and the Standard Deviation to Describe a Set of Data?

The mean is the one most representative value to describe a set of scores, and the standard deviation is the most representative value to describe the spread or variability of a set of scores.

These two descriptive measures can be used together to describe a set of scores in which, for example, the mean for a set of test scores is 100 and the standard deviation is 10.

For this example, given the characteristics known about the bell-shaped curve, the following diagram shows how the data can be described.

A Normal Curve Divided Into Different Sections

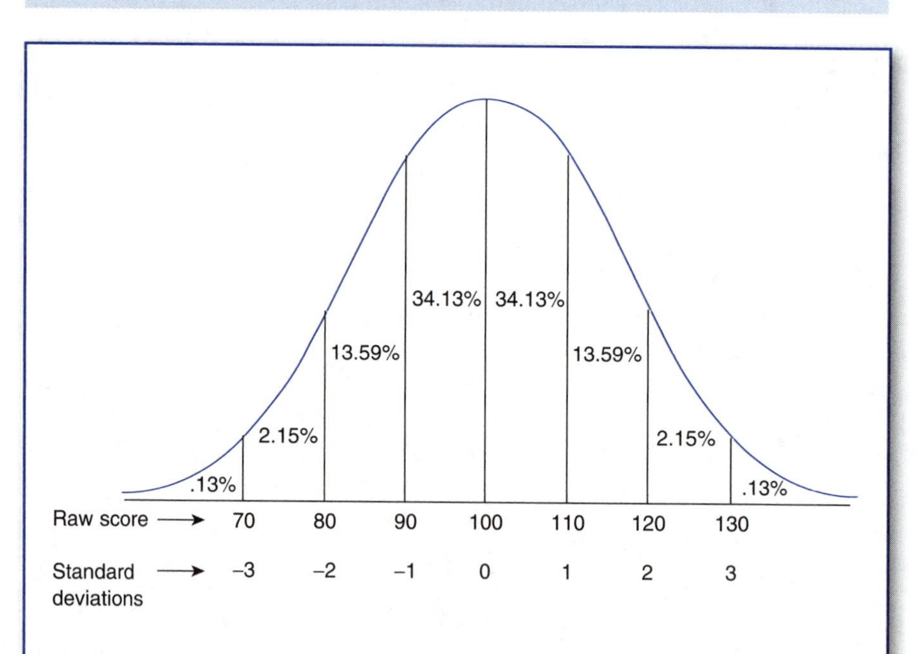

- 68% of all scores fall between +1 and −1 standard deviations from the mean, or, in this example, between scores of 90 and 110.
- 95% of all scores fall between +2 and −2 standard deviations from the mean, or, in this example, between scores 80 and 120.
- 99% of all scores fall between +3 and −3 standard deviations from the mean, or, in this example, between scores of 70 and 130.

The above diagram allows a fairly accurate estimate of the likelihood of any one score in the set of scores. For example, there is a 50% chance of a score appearing below 100 (the centermost score). Or, there is a 64% chance of a score appearing between 90 and 110 and a 5% chance of a score appearing above 120.

Being able to estimate probabilities associated with scores (as I just showed) becomes a very powerful tool for helping to make decisions about particular outcomes. For example, what is the likelihood that any one score "fits" in a set of scores? The example given tells us that it is pretty likely (greater than 95%) that a score of 120 fits in this set of scores.

This model can then be used to estimate how unique a particular outcome might be and, if unique enough, be termed significantly different from what would be expected by chance alone.

More questions? See #50, #55, and #59.

QUESTION #55

What Is a Normal Curve, and What Are Its Characteristics?

A normal curve, sometimes called a bell-shaped curve, is a distribution of scores that forms the basis for much of our understanding and use of descriptive and inferential statistics. It allows researchers to estimate the probability that is associated with an outcome.

As you can see in the diagram below, a normal curve has three characteristics.

The Normal, or Bell-Shaped, Curve

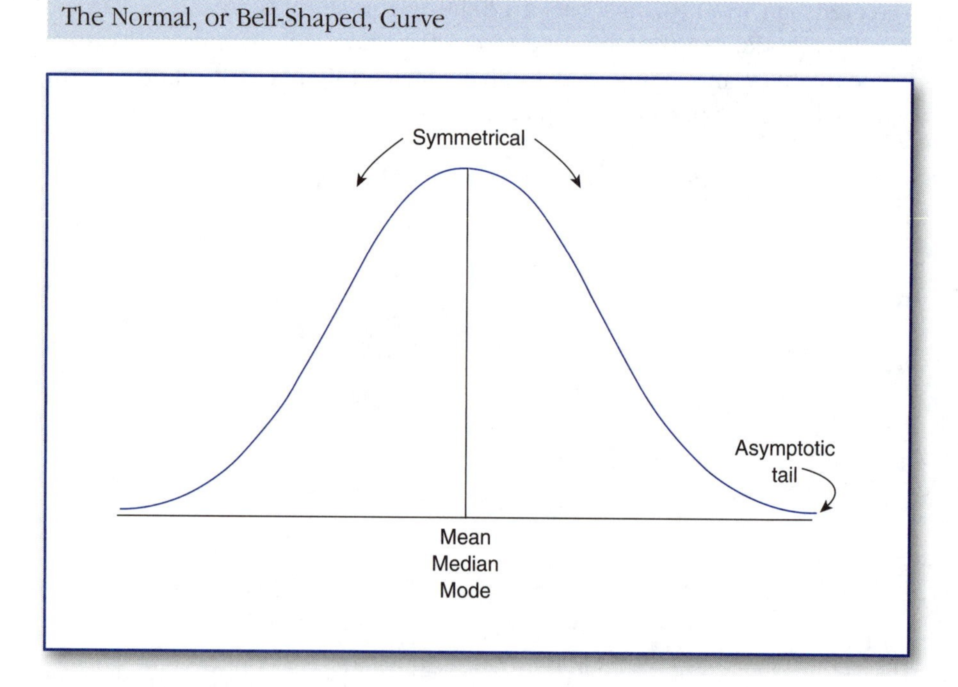

1. It is symmetrical about the mean, meaning that each half of the curve is a mirror image of the other. Each half "contains" 50% of the scores in the entire set of scores.

2. Its tails are asymptotic, meaning that they never meet the x-axis and that the likelihood associated with a specific outcome decreases as

the outcome becomes more extreme. As you move toward the highest of values and the lowest, the likelihood of those values occurring is less than the more moderate scores. And, no matter how extreme an outcome, there is always a probability associated with it, even if that probability is very small (which is why the tails never touch the axis).

3. In the normal curve, the mean, the mode, and the median are all equal to one another.

Why is the normal curve such an important part of understanding statistics? It's a framework or the model for what we would expect by chance for particular outcomes.

When a statistical test is conducted, the results of that test are compared to a distribution of scores (such as the normal curve but sometimes other distributions). The results of the test can then be compared with what would be expected by chance, and if the difference between the two is of sufficient magnitude (and there are tables and other techniques to determine such), then the outcome is deemed unique or significant. The normal curve provides us with a standard to make that decision.

More questions? See #50, #56, and #59.

QUESTION #56

If a Distribution of Scores Is Not Normal (or Not Bell Shaped), How Can the Ideas on Which Inference Is Based Be Applied?

You can imagine how often researchers have to deal with populations of data in which the distribution of scores is not normal or bell shaped. And, much of what researchers discuss is based on the notion that such populations are normally distributed. So, how does one know if the population distribution from which a sample is selected is normal? The answer is that you don't, because you can never actually examine or evaluate the characteristics of the entire population. And, because of the central limit theorem, it does not matter.

The central limit theorem says that regardless of the shape of the population (be it normal or not), the means of all the samples selected from the population will be normally distributed. This means that even if a population of scores is shaped like a U (the exact opposite of a bell-shaped curve), if you select a number of samples of size 30 or larger from that population, the means of those samples will be normally distributed. Most important, nothing about the distribution of scores within the population needs to be known to generalize results from the sample to the population.

This theorem is important stuff. It illustrates how powerful inferential statistics can be in allowing decisions to be based on the characteristics of a normal curve when, indeed, the population from which the sample was drawn is not normal. This fact alone provides enormous flexibility and in many ways is the cornerstone of the experimental method. Without the power to infer, the entire population would have to be tested—an unreasonable and impractical task.

More questions? See #55, #57, and #59.

QUESTION #57

What Does It Mean When a Distribution Is Skewed?

When a distribution is skewed, it means that it is not normal or it is not bell shaped. So scores can be positively skewed or negatively skewed.

A positively skewed set of scores (sometimes referred to as left skewed) looks like the figure you see below. It represents scores where the right tail of the distribution is longer and there is an overconcentration of scores on the left-hand side of the distribution. When a distribution is positively skewed, the mean, median, and mode also appear as shown in the accompanying figure.

A negatively skewed set of scores (sometimes referred to as right skewed) looks like the figure below. It represents scores where the left tail of the distribution is longer and there is an overconcentration of scores on the right-hand side of the distribution. When a distribution is negatively skewed, the mean, median, and mode also appear as shown in the accompanying diagram.

Degree of Skewness in Different Distributions

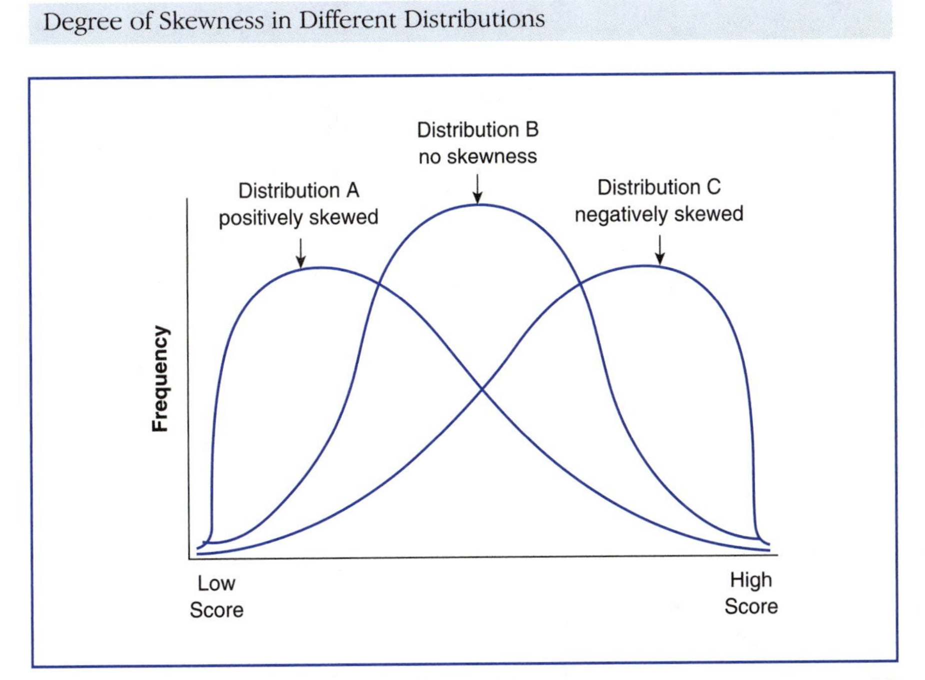

There are some common measures of skewness that you may encounter in your studies and reading of the literature. One of the most commonly used and most simple was that proposed by Pearson, where the formula is . . .

$$\text{Skewness} = (\text{Mean} - \text{Mode})/\text{Standard Deviation}$$

So, for example, if the mean equals 50, the mode equals 10, and the standard deviation equals 5, the skewness coefficient equals 8. As the mean and the mode become further apart (and skewness increases), the skewness coefficient increases, and this can be a negative or a positive number depending upon the relationship between the mean and the mode. For example, a mean of 50 and a mode of 5 with a standard deviation of 5 results in a skewness coefficient of 9.

More questions? See #55, #56, and #58.

QUESTION #58

I'm Looking for a Visual Way to Describe Data. What Are Some of My Choices?

A picture is easily worth at least a thousand words, and there are a variety of ways that you can summarize and illustrate data to help communicate outcomes.

One of the most common ways is by using a frequency distribution, which tallies the number of occurrences within a series of class intervals. This is a simple practice of tallying individual occurrences of scores (and their total frequency) for a band of values such as you see here for a total of 100 scores.

Class Interval	Frequency
45–49	3
40–44	8
35–39	8
30–34	22
25–29	25
20–24	13
15–19	12
10–14	4
5–9	4
0–4	1

If you take a frequency distribution and create a line graph of the data, you have a frequency polygon, which is a continuous line that represents the frequencies of scores within a class interval, as shown here.

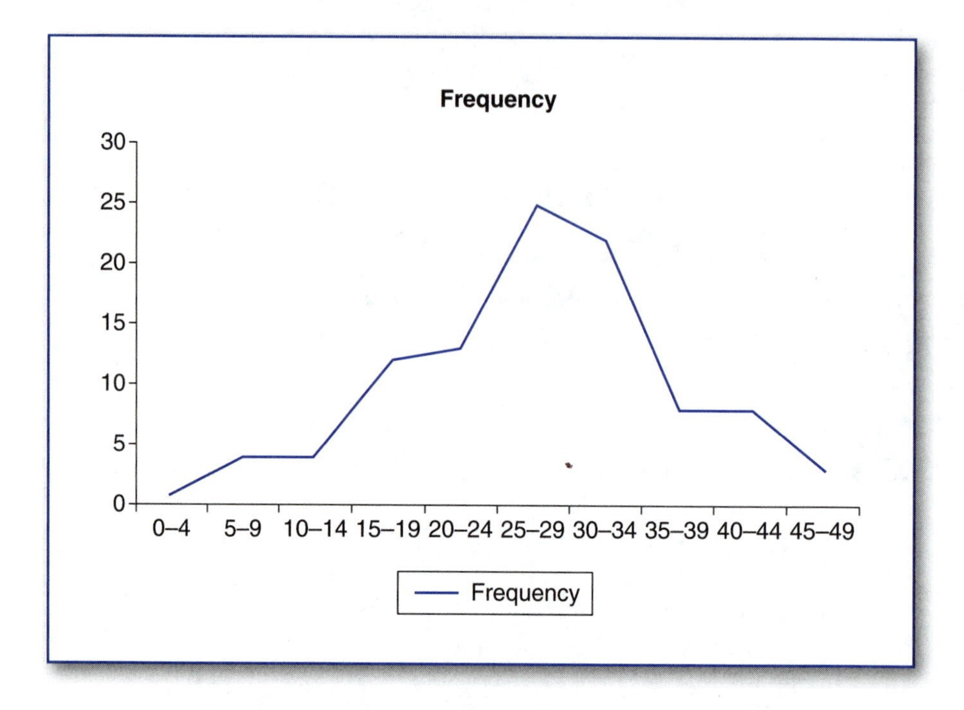

Here you can see the frequency of a certain class of scores as a function of the category of scores and examine the nature of the distribution. You can even speculate on such descriptive statistics as the mean and how much variability might characterize the distribution.

More questions? See #50, #55, and #57.

QUESTION #59

What Is a Standard Score, and Why Is It Important?

A standard score is a value that is computed using the standard deviation as units of distance from the mean. The most important thing about a standard score is that, once computed, it allows for the comparison of raw scores from different distributions. What this means, in effect, is that regardless of the shape of a distribution (no matter what its mean, standard deviation, or degree of skewness), a standard score can be computed and can be directly compared to another standard score from a distribution that has entirely different characteristics.

The following table illustrates this point . . .

Sample	Mean	Standard Deviation	Raw Score	Z Score
A	100	10	100	0
B	100	5	90	1
C	100	10	90	2
D	50	4	54	1

As you can see for Sample A, when the mean and the raw score are equal, the deviation from the mean is zero and the standard score is zero as well. This illustrates that any z score of zero means that the raw score and the mean are equal.

In Sample B, a raw score of 90 has a standard score of 1 because it is one unit of standard deviation removed from the mean.

In Sample C, a raw score of 90 has a standard score of 2 because it is two units of standard deviation removed from the mean.

For Sample D, it is a much different situation. Here, the mean, standard deviation, and raw score clearly come from a very different distribution with a standard score of 1 representing the raw score of 54. Yet, a raw score of 54 and a raw score of 90 (Sample B) are directly comparable.

Standard scores are also used to create certain distributions (such as in the use of the Z test) that are used in inferential statistics.

More questions? See #55, #56, and #60.

QUESTION #60

What Are Some of the More Common Standard Scores, and How Are They Used?

The most common types of standard scores are z scores and T scores, both standard scores that are frequently reported in research studies. Both are computed using the mean and standard deviation of the respective distributions.

z scores are computed using the following formula . . .

$$z = \frac{X - \bar{X}}{s}$$

where

X represents the raw score,

$\bar{X}$ represents the mean of the sample, and

s represents the standard deviation of the sample.

For example, for a sample with a mean of 100 and a standard deviation of 10, a raw score of 5 would have a z score of 2. This score is 2 standard deviations above the mean.

T scores are computed using the following formula . . .

$$T = z + 50$$

For example, for a sample with a mean of 100 and a standard deviation of 10, a raw score of 5 would have a T score of 52. The primary purpose of using a T score rather than a z score is to eliminate negative values.

Both of these standard scores serve the same purpose—to be able to compare scores from different distributions to one another. So, regardless of whether a score in a sample distribution is 41 or 1,900, if they both have T scores of, say, 40 (which would mean that the raw scores were below the mean of the samples), they are directly comparable to *T* scores from other distributions. Similarly, a raw score of 75 from two different distributions may not be equal at all because the average or standard deviation of both distributions may differ markedly.

The fact that standard scores are directly comparable makes cross-distribution comparisons easy and practical. It's a way to compare scores to one another when they are from very different distributions.

More questions? See #53, #55, and #59.

ALL ABOUT TESTING AND MEASURING

QUESTION #61

There Is a Particular Outcome That I Want to Measure, But I Have No Idea Where I Can Find Out Whether or Not There Are Existing Measures. Where Do I Look to Find Suggestions as to What Dependent Variable I Should Use?

At this point, your job is to find a measure that best reflects the central idea behind your research activity, and there are many different places to look. Perhaps the best is to do a library search of journal publications that have examined the same research question (or a variant thereof) as you are proposing. Find out what measures they have used and then make a judgment as to whether they are right for you.

What other criteria might you use to make such a judgment? Here's are some questions worth considering . . .

1. Is the measure both reliable and valid? These are perhaps the two most important criteria.
2. Am I qualified to administer it? Some outcome measures require certain training or skills to administer, and you may not have the time or money for such training.
3. Can I easily obtain a copy of the instrument?
4. Is it available only to experienced researchers, or will I be able to contact the company that publishes the measuring tool or one of the authors of the tool?
5. Does it accurately assess what I need to know?
6. Will it work in the research setting in which I intend to use it?

Once you have consulted the library literature and have a good idea whether what you need is available, you may want to also explore what The Buros Institute of Mental Measurements, now at the University of Nebraska, has to offer. You can find them at http://www.unl.edu/buros/, and they have more than 3,500 reviews of existing outcome variables and tests. This resource is available through your university, college, or community library and also in the print edition of the *Eighteenth Mental Measurements Yearbook*.

More questions? See #63, #66, and #71.

QUESTION #62

What Are the Different Levels of Measurement, and How Are They Used?

There are four levels of measurement, each one increasing in precision from the nominal through the ratio level.

The nominal level of measurement characterizes variables that are distinguished in name only (e.g., hair color, with groups such as black, blond, and brown).

The ordinal level of measurement characterizes variables that are distinguished by their magnitude (e.g., class rank, with groups such as higher and lower).

The interval level characterizes variables that are distinguished by their magnitude and a scale along which there are equally appearing intervals (e.g., test scores, with scores such as 82 correct and 90 correct).

The ratio level of measurement characterizes variables that are distinguished by having an absolute zero, where the presence of the trait or characteristic is not present.

Here are some things that researchers need to remember about scales of measurement.

1. Any variable that is assessed can be assigned to a certain level of measurement.
2. How a variable is measured determines its level of measurement. For example, one can organize children based on height into different groups of tall and short children, which is ordinal in nature. Or, children can be categorized by height in inches, which is at least interval.
3. The "higher" up the scale of measurement (with nominal being lowest and ratio being highest), the more precise the level of measurement. Knowing that someone scored 10 out of 12 on a measure is much more precise, accurate, and informative than knowing someone is better at spelling than others.
4. Variables at different scales of measurement correspond to different descriptive and other analytic techniques.

More questions? See #61, #68, and #79.

QUESTION #63

What Is Reliability?

Reliability is the quality of a test or assessment tool that says it is consistent. Now, it may be consistent over time or across different forms of the test.

Once a test is administered, the primary result is a score. We'll call that the *observed score* because it is what's observed and is the only actual indication of performance available. So if Susan gets 78% of the questions correct on her English literature midterm, that's the observed score.

Any observed score is made up of two different types of other scores—true score and error score—so the entire idea looks something like this . . .

Observed Score = True Score + Error Score

True score is defined as the absolutely correct score, yet it is a theoretical concept and is unobservable. It's what we would count as an absolute and unequivocal reflection of Susan's knowledge.

Error score is what accounts for the difference between observed score and true score. For example, if a student is sick on the day of a test, we can assume that the illness may interfere with his performance and that his observed score would not match his true score.

There are two kinds of errors that can occur. One is trait error, which has to do with the individual (sick, unprepared, tired, bored, or unmotivated), and method error (poor lighting or poor instructions).

If one can minimize the amount of error score that enters into the equation, then the observed score more closely matches the true score. When both of these are equal, we can then talk about a test being perfectly reliable. And, the farther apart these values are from one another, the less likely the test is to be reliable because, over time or across forms, the individual's observed score does not accurately reflect the individual's true score.

More questions? See #64, #65, and #71.

QUESTION #64

What Are Some of the Different Types of Reliability, and When Are They Used?

There are generally four types of reliability, all computed slightly differently, but all characterized by the idea that reducing uncertainty or error, and explaining more true scores, increases reliability.

The following table provides an overview of these different types and a general answer to the important question regarding each type.

Type of Reliability	What It Does	How It Is Computed	When It Is Used	An Example
Test-Retest Reliability	Computes the reliability of the same test administered at two different times.	A correlation is computed between scores at Time 1 and scores at Time 2 for the same individuals.	When a researcher is interested in whether the instrument under development is reliable from one point in time to another.	A researcher is interested in knowing whether a test is consistent when administered at the beginning of a program and then administered 5 months later when the program ends.
Parallel Forms Reliability	Computes the reliability of two different tests that measure the same outcome administered closely together in time.	A correlation is computed between scores for Form 1 and Form 2.	When a researcher needs to use different forms of the same test at the same time.	A researcher is examining memory skills and needs to have participants be tested on two different recall tasks that both measure the same things but have different items.

Type of Reliability	What It Does	How It Is Computed	When It Is Used	An Example
Internal Consistency Reliability	Computes the extent to which a test is unidimensional and measures only one construct.	A correlation is computed between individual item scores and total test score.	When a researcher needs to know if the test under construction assesses one or more dimensions.	A researcher is developing an intelligence test and wants to make sure that each subtest is internally consistent and assesses only one dimension.
Interrater Reliability	Looks at the percent of agreement between two or more raters.	Interrater reliability is a measure of the number of times raters agree divided by the total number of possible agreements.	When two or more raters are evaluating or assigning values to observations.	Two researchers are evaluating the use of an observation tool aimed at classifying supermarket buying behavior into five categories.

More questions? See #63, #65, and #71.

QUESTION #65

How Are Reliability Coefficients Interpreted?

Regardless of the type of reliability coefficient that you are dealing with (test-retest, internal consistency, etc.), you want to be able to interpret the importance of its size.

"Good" reliability is indicated by two qualities of correlation coefficients.

First, you want reliability coefficients that are positive. A negative reliability coefficient indicates that there is something greatly amiss with the instrument.

Second, you want reliability coefficients that are as large as possible.

For example, here are three types of reliability coefficients along with how they might be interpreted and next steps to be taken (if any).

Type of Reliability	Sample Value	Interpretation	What's Next?
Test-retest reliability	.78	The test is reasonably consistent over time. A reasonable goal is for the coefficient to be above .70, and, better, to be in the .80s or .90s.	Not much. This is a pretty reliable test, and it can be administered with confidence.
Parallel forms reliability	.17	The test does not seem to be very consistent over different forms. The value .17 is a very low reliability coefficient.	Work on the development of a new and better form of the test.
Internal consistency reliability	.38	The test does not seem to be uni- or one-dimensional in that these items are not consistently measuring the same thing.	Be sure that the items on the test measure what they are supposed to (which, by the way, is as much a validity issue as a reliability issue—stay tuned for the next chapter).

In general, an acceptable reliability coefficient is .70 or above, but much more acceptable is .8 and above. When it comes to interrater reliability, you should really expect nothing less than 90%. It's so easily raised (just have the judges do more training, given that the terms on the test are good ones) that there is no reason why this higher level should not be reached.

More questions? See #63, #64, and #71.

QUESTION #66

What Are Some of the Different Types of Validity, and When Are They Used?

Validity is the quality of an assessment tool that tells the researcher whether it does what it claims to do. For example, one would expect a spelling test to assess spelling skills and a test of spatial relations to assess spatial relations. If the test can be used to assess this variable, it is valid. A test cannot be valid if it is not reliable because only when a test can measure something reliably (time and again) can it accurately measure what it is designed to.

There are several types of validity as follows.

Content validity is used when you want to know whether a sample of items truly reflects an entire universe of items within a certain topic. An example of content validity is a history test on a Civil War unit or an Advanced Placement physics test.

Criterion validity is used when you want to know if test scores are systematically related to other criteria that indicate the test taker is competent in a certain area. Criterion validity can take the form of concurrent or predictive validity, and an example is the EATS test (of culinary skills taken upon graduation) being correlated with a chef's reputation 5 years after graduation from culinary school.

Construct validity is used when you want to know if a test measures some underlying psychological construct. An example might be looking at how well a new test of masculinity is correlated with those traits or behaviors that theoretically define the construct. Construct validity is the most ambitious and time consuming to establish because it deals with related complex social and behavioral groups of behaviors.

More questions? See #67, #69, and #72.

QUESTION #67

What Is Criterion Validity, and How Do the Two Types of Criterion Validity, Concurrent and Predictive, Differ?

Criterion validity assesses whether a test reflects a set of abilities in a current (concurrent validity) or future (predictive validity) setting as measured by some other test (the criterion). Criterion validity is used most often for achievement tests and certification or licensing. In both cases, a criterion or some outside measure is used to establish validity. Both concurrent and predictive validity are established using simple correlations where scores on the instrument under development are correlated with the criterion.

The key to establishing either concurrent or predictive validity is the quality of the criterion. If the criterion is not an accurate reflection of what needs to be measured, then the correlations are meaningless.

For example, to establish the predictive validity of a school entrance exam, the following simple steps might be taken.

1. The school entrance exam would be created and administered.
2. School entrance exam scores would then be correlated with later high school graduation rates. If they are correlated, it could mean that the school entrance exam has predictive validity.

Or, to establish the concurrent validity of a child care exam, the following simple steps might be taken.

1. The child care exam would be created and administered.
2. Child care professionals would observe child care students and rank them. Child care exam scores would then be correlated with the observers' rankings. If they are correlated, it could mean that the child care exam has concurrent validity.

More questions? See #66, #70, and #71.

QUESTION #68

What Is the Difference Between a Norm-Referenced and a Criterion-Referenced Test?

Although achievement tests all sample a universe of knowledge, they can take different forms as far as the reference points used to assess the success of the outcome of the test.

Achievement tests are most often norm-referenced or criterion-referenced.

Norm-referenced tests allow a comparison of an individual's test performance to the test performance of other individuals. For example, if an 8-year-old student receives a score of 56 on a mathematics test, the norms that are (or may be) supplied with the test can be used to determine that child's placement relative to other 8-year-olds. In other words, did the test taker do better or worse than others who have taken the test? Standardized tests are usually accompanied by norms.

Criterion-referenced tests (a term coined by psychologist Robert Glaser in 1963) define a specific criterion or level of performance, and the only thing of importance is the individual's performance, regardless of where that performance might stand in comparison with others. In this case, performance is defined as a function of mastery of some content domain. For example, if you were to specify a set of objectives for 12th-grade history and specify that students must show command of 90% of those objectives to pass, then you would be implying that the criterion is 90% mastery.

Some criterion-referenced tests are also referred to as high-stakes tests. Here, the results of the test have important consequences for the individuals who take the test. Tests for admission to professional schools such as business and medicine could be considered high-stakes tests. Because this type of test actually focuses on the mastery of content at a specific level, it is also referred to as a content-referenced test.

More questions? See #62, #75, and #79.

QUESTION #69

What Is Construct Validity, and Why Is It Especially Appropriate for Establishing the Validity of Psychological Tests?

Construct validity is the quality of a test such that the assessment tool accurately measures some underlying construct. A construct is a set of interrelated variables such as attention, aggression, and attachment. All of these constructs consist of a group of variables that is unified by some underlying theory or theories.

For example, attachment is often thought of as a construct because it can be viewed as a group of interrelated variables such as touch, verbal behavior, and smiling. Tests that purport to measure such constructs establish their validity in one of several ways.

The first is by correlating the scores on the new test with scores from participants who have taken a similar test in the recent past. A simple correlation is used for this, and the idea is that if one test is shown to be valid, then any that are created in a similar fashion will be as well.

Another method, and a much more ambitious one, is known as the multitrait/multimethod, where several different traits are measured using several different methods (including the test for which construct validity is being sought). All of these variables are correlated, and the researcher looks for evidence that similar traits are related regardless of the method used for measurement and that different traits are unrelated using the same measurement tools. This is a complex but very effective way to establish construct validity.

More questions? See #66, #70, and #71.

QUESTION #70

How Are Different Types of Validity Established?

Content, criterion, and construct validity are all established in different ways (although the same techniques may be used). Here's a chart that summarizes the type of validity, the type of test for which it is used, and how it is established.

Type of Validity	Used to Establish the Validity of . . .	Method Used to Establish Validity
Content	achievement tests	A content expert examines the items and makes a decision as to whether the test items represent the universe of all possible test items. For example, does a physics test on the laws of thermodynamics contain items that represent all the possible questions about that topic?
Criterion	employment and vocational tests	For concurrent validity, the test is correlated with an established measure of the desired behavior, and both are done within the same time period. For example, how well do scores on a test of mechanical skills correlate with scores on assembling a three-dimensional model of an engine? For predictive validity, the test is correlated with an established measure of the desired behavior, which takes place in the future. For example, how well do scores on a test of mechanical skills correlate with future success as an architect as measured by number of clients?

Type of Validity	Used to Establish the Validity of . . .	Method Used to Establish Validity
Construct	psychological traits and tests of certain characteristics	One method to test for construct validity is to assess underlying traits measured by already valid instruments and correlate those scores with the new items. A second method is to establish multitrait/multimethod validity, in which scores using the same methods and measuring different traits are uncorrelated, whereas scores using different methods and measuring the same traits are correlated.

More questions? See #63, #65, and #71.

QUESTION #71

How Do Reliability and Validity Work Together?

You remember that reliability is defined as the quality of an assessment tool or a test that shows it to be consistent over time and over different forms. In other words, Jim will receive the same score on a test (relative to his classmates) each time he takes it.

And, validity is the quality of a test that says it does what it is supposed to do. In other words, a sophomore test of events that led up to the Civil War actually does test exactly that and nothing else.

The relationship between validity and reliability is a fairly simple, but interesting one. For a test to be valid, it must first be reliable. Said another way, a test must be consistent and do the same thing time after time, before one can be sure that it does what it is supposed to do.

For example, let's take a very simple test question and assume that this is one of 50 such items, all different, but on the same topic and of the same kind of items from a highly reliable test.

1. Which of the following is not a type of reliability?
   a. Test-retest
   b. Parallel forms
   c. Content analysis
   d. Internal consistency

This item could easily belong on a test for a course in tests and measurement, but what if you were told it is an item from a history test? This may be an item from a highly reliable test, where the test is consistent over time and forms, but surely not a valid measure of history.

In sum . . .

1. Tests have to be reliable before they can be valid.
2. A test can be reliable but not necessarily valid.
3. A test cannot be valid without being reliable.

Classroom tests, personality tests, and any other kinds of tests need to be reliable before they can be deemed valid. That's how they relate to each other and why they are so critical to one another.

More questions? See #63, #66, and #70.

QUESTION #72

How Can I Find Out If a Test Is Reliable and Valid?

There are two very good ways to find out if a test is reliable and valid as first steps in determining if it meets your needs and the needs of your project. First, seek out studies that have already used the test in published (and refereed) research. The second best way is to consult with the publisher or author of the test and simply ask for the data showing that the test works.

Also, keep the following in mind.

First, if the test has a long and established record of being used and being popular by way of its psychometric qualities, then you can assume that it is a "good" test and one you can use without any further concern. How you know that the test has a long and established record is a result of your reading and reviewing the literature.

Second, if the test is relatively new and has not had a great deal of exposure to the research audience, look for some comment on the part of the authors in a research report about the qualities of the test. They should mention reliability and validity data and any other important qualities, uses, or issues that may be pertinent to your use of the test as an important variable.

And, if there is no mention whatsoever of the reliability and validity qualities of a test, and if you have not seen it reviewed or mentioned in other literature, then be somewhat skeptical as to whether it is ready to be used in your own research.

Most often, if the qualities are not mentioned, then the quality is not there. Researchers who go through the trouble to clearly document the psychometric properties of a test, and those properties indeed indicate the test is useful, will surely mention it in their report or journal article.

More questions? See #71, #73, and #79.

QUESTION #73

What Are Some of the Different Types of Tests, and How Are They Used?

There are many different types of tests, and they each have a specific purpose. Here's a brief table that should acquaint you with the main types and what they do, including an example.

Type of Test	What It Does	An Example
Achievement	Achievement tests measure knowledge of a particular subject area and previous learning.	• EXPLORE • Fast Track • Iowa Tests of Basic Skills
Personality	Personality tests measure an individual's enduring characteristics and disposition.	• Adolescent Psychopathology Scale • Family Relationship Inventory • Multidimensional Self Concept Scale
Aptitude	Aptitude tests measure the potential to learn or acquire a skill.	• Evaluation Aptitude Test • Senior Aptitude Tests • Trade Aptitude Test Battery
Intelligence	Ability tests measure intellectual potential (and intelligence tests fall in this area).	• Comprehensive Test of Nonverbal Intelligence • Wechsler Abbreviated Scale of Intelligence • Wide Range Intelligence Test
Career or Vocational	Career choice or vocational tests determine levels of interest in various occupations and the skills that are associated with those occupations.	• Accountant Self Selector • Career Survey • Fine Dexterity Test

Type of Test	What It Does	An Example
Education	Education tests are general tools that assess instructional/school environment, effective teaching, learning styles, and educational leadership.	• CLEP Exam in Educational Psychology • Learning Preference Scales • The School Leadership Series
Fine Arts	Fine arts tests assess knowledge, skills, and interests in fine and performing arts as well as tests of aesthetic judgment.	• Advanced Placement in Examination in Studio Art • Instrument Timbre Preference Test • Iowa Tests of Music Literacy

More questions? See #71, #78, and #79.

QUESTION #74

When It Comes to Measuring Attitude, What Is the Difference Between a Likert and a Thurstone Scale?

Attitude is an individual's feelings about a person, idea, object, or event. If you like something, that's an expression of attitude, and there are generally two methods used to measure attitude.

The first is a Likert scale, also referred to as the method of summated ratings, which results in a 5- to 7-point scale that looks something like this . . .

——+——	——+——	——+——	——+——	——+——
Strongly Agree	Agree	Undecided	Strongly Agree	Agree

Test takers' responses to items such as "Chocolate should be a part of every meal" and individual scores are a function of the sum of all items. Likert scales are developed by writing out a collection of items that express a feeling or option; items that express a stronger sentiment are also included, as well as items to which test takers can respond. Items are often "reversed" in sentiment (such as "Chocolate should be a part of every meal") to ensure that a test taker does not just arbitrarily mark one or the other extreme for each item.

Thurstone scales, also referred to as equally appearing interval scales, are created by collecting a large number of items that express some sentiment, having individuals act as judges to place the items somewhere along a scale ranging from extremely unfavorable to extremely favorable (or some other labels that express extremes), and then evaluating across judges which items are least variable. The idea is that consensus produces items with a certain value attached to them, and that value is the score on each item for any one individual.

Which to use? Check and see if an attitude scale already exists for the topic in which you are interested, and if not, either one can be equally effective.

More questions? See #62, #72, and #73.

QUESTION #75

What Is Item Analysis, and How Is It Used in Evaluating Achievement Tests?

Achievement tests, such as those administered in classrooms, focus on assessing whether an individual has knowledge of a certain content. Often, items on such tests are multiple choice in nature and contain a stem and alternatives.

For example, here's the stem . . .

Item 9. The sum of 5 + 10 =

And here are some alternatives . . .

A. 10
B. 25
C. 15
D. 5

The job of item analysis is to analyze all the responses to each item and tell you (as the teacher or the test designer or the test administrator) whether the items are "good" or not (and more about that shortly).

An item analysis of an item consists of generating two scores for each item.

The first is the difficulty index, and this is the percent of individuals who got the item correct. It ranges from 0% (no one got it correct) to 100% (everyone got it correct).

The second index is discrimination, which is the percent of test takers in the higher scoring range who got the item correct as opposed to the percent of test takers in the lower scoring range who got the item correct.

These two indexes work in tandem with each other to help the test creator produce the best items.

So, what is a good item? The best items are those that perfectly discriminate between those who did well on the test and those who did not do well. In other words, the people who scored high on the test got the item correct and those who scored low on the test did not. If this were the case, the discrimination index would be 1 or 100%.

As far as difficulty, the best items are those for which the difficulty level is 50%—half of all the test takers got the item correct and half got it incorrect. Which halves? The ones who got the item correct are the high scorers and the ones who got the item incorrect are the low scorers. In fact, item discrimination is restricted by item difficulty. It is only when 50% of the test takers get the item correct that the item can fully discriminate.

More questions? See #68, #72, and #73.

QUESTION #76

What Is a Percentile or a Percentile Rank?

Raw scores and transformed or standard scores yield different types of information, and we have already reviewed how standard scores are important because they allow us to compare scores from different distributions to one another.

Another way of understanding individual raw scores is to examine the score relative to the rest of the scores in the data set. For example, we know that an individual got a score of 98 on an achievement test, but what does that mean relative to the other scores in the group?

A percentile or a percentile rank is a point in a distribution of scores below which a given percentage of scores fall. It's a particular point that falls within an entire distribution of scores, and there are 99 of them (there can't be 100 because the highest score cannot exceed itself).

The 45th percentile, for example, is the score below which 45% of all the other scores, in the total set of scores, fall. Percentiles and percentile ranks (terms that are often used interchangeably) are probably the most often-used score for reporting test results. When percentiles or percentile ranks are reported, they often looks like this: P_{45}.

But you may have also noticed that this definition of a percentile tells us nothing about absolute performance. If someone has a percentile rank of 45, it means that he or she could have a raw score of 88 or a raw score of 22—we just don't know. However, the lower the percentile, the lower the person's rank in the group.

More questions? See #62, #73, and #75.

QUESTION #77

What Is Adaptive Testing?

Every testing instrument has as the primary goal getting as close as possible to the true score of whatever is being measured. A relatively new model of testing is adaptive testing or Computer Adaptive Testing. Adaptive testing adjusts the difficulty of items based on the test taker's ongoing performance. In other words, based on the previous items that the test taker has gotten right or wrong, the program will choose subsequent items to maximize efficiency and get closer to the test taker's true score. Adaptive testing accomplishes this through a series of steps.

- First, the test taker begins taking the test. Where he or she starts is a function of the test giver's estimate of the test taker's ability. Because the process is computerized, the items can be presented according to the test taker's performance.
- Second, all the test items in the test bank are evaluated to determine which one is the best next one given the test taker's performance up to that point.
- Third, the best next item is administered.
- Fourth, the computer program estimates a new ability level based on the test taker's responses to all of the administered items.

Steps 2 through 4 are repeated until the test ends.

The goal is to get as close as possible to the test taker's true score, thereby producing a more reliable and more valid assessment. A list of tests that are currently available as adaptive tests as well as information about the adaptive testing can be found at C.A.T. Central at http://www.psych.umn.edu/psylabs/catcentral/.

Advantages to adaptive testing are more precision, less time because the same level of accuracy can be reached with 50% of the items, test takers waste little time on items that are too easy or too hard, and results are immediately available.

More questions? See #63, #73, and #78.

QUESTION #78

What Is the FairTest Movement, and What Are Its Basic Goals?

Testing is big business, and lots of people see it as placing the test taker at an unfair advantage because it has become such a huge industry with sometimes more emphasis on collecting scores than accurately evaluating performance and using those results in an effective way.

The FairTest movement was created to help ensure that "evaluation of students, teachers and schools is fair, open, valid and educationally beneficial" (http://www.fairtest.org/). Here are FairTest's basic principles and their exact words are in italics.

1. *Assessments should be fair and valid.* They should provide equal opportunity to measure what students know and can do, without bias against individuals on the basis of race, ethnicity, gender, income level, learning style, disability, or limited English proficiency status.
2. *Assessments should be open.* The public should have greater access to tests and testing data, including evidence of validity and reliability. Where assessments have significant consequences, tests and test results should be open to parents, educators, and students.
3. *Tests should be used appropriately.* Safeguards must be established to ensure that standardized test scores are not the sole criterion by which major educational decisions are made, and that curricula are not driven by standardized testing.
4. *Evaluation of students and schools should consist of multiple types of assessment conducted over time.* No one measure can or should define a person's knowledge, worth, or academic achievement, nor can it provide for an adequate evaluation of an institution.
5. *Alternative assessments should be used.* Methods of evaluation that fairly and accurately diagnose the strengths and weaknesses of students and programs need to be designed and implemented with sufficient professional development for educators to use them well.

More questions? See #20, #73, and #77.

QUESTION #79

Where Do I Find a Collection of Tests From Which to Choose? And, How Do I Go About Selecting One?

When it comes time to select a test to be used in your research, there are several places that are a good place to start. These resources should be a supplement to an exhaustive search of the literature and a review of articles that focus on the same topics as your work. You can see which outcome measures were used and then review those.

For online resources, the best collection of information about tests and reviews is the Buros Institute of Mental Measurements at the University of Nebraska (http://www.unl.edu/buros). Here you can find reviews of tests in the following 18 categories.

- Achievement
- Behavior Assessment
- Developmental
- Education
- English and Language
- Fine Arts
- Foreign Languages
- Intelligence and General Aptitude
- Mathematics
- Miscellaneous
- Neuropsychological
- Personality
- Reading
- Science
- Sensory-Motor
- Social Studies
- Speech and Hearing
- Vocations

You can also search alphabetically for the name and associated reviews of a particular test. The tests themselves are not available, and you have to contact the publisher of the test to gain permission to use it as well as to obtain copies.

A sample review consists of the following information: title, purpose, population, publication date, acronym, score, group or individual administration, price data, time for administration, authors and publisher, plus an extensive review of the test itself done by a qualified expert in this substantive area. Although the Buros site is free to access and get some information, access to the more than 3,500 reviews is not free, and the reviews cost about $15 each. Note that the Buros collection is often available through college and university websites, costing the registered user nothing.

More questions? See #66, #68, and #72.

PART 8

UNDERSTANDING DIFFERENT RESEARCH METHODS

QUESTION #80

What Is an Experimental Design, and What Is the Difference Between the Major Types of Experimental Designs?

An experimental design is the plan you use to answer the cause-and-effect research question you are asking and to test the relevant research hypothesis. It's the grand plan, and there are basically three types: pre-experimental, quasi-experimental, and true experimental. Here's how they differ from one another . . .

	Pre-experimental	**Quasi-experimental**	**True Experimental**
Is there a control group present?	Usually not	Often	Yes
Are participants randomly selected from a population?	No	Yes	Yes
Are participants randomly assigned to groups?	No	Yes	Yes
Are treatments randomly assigned to groups?	No	Yes	Yes
Amount of precision and control	Least	Some	Most

Here is an example of a very simple pre-experimental design called a one-shot case study.

	Pretest	Treatment	Posttest
Experimental Group	No	Yes	No

For example, the nutritional habits of a 90-year-old resident of a graduated care independent living home is assessed through the creation and completion of a daily diary. The "treatment" is the collection of the data, but there is no comparison with other individuals or groups.

Here is an example of a very simple true experimental design called a post-only control group design.

	Pretest	Treatment	Posttest
Experimental Group	No	Yes	Yes
Control Group	No	No	Yes

In such a design, the experimental and the control group are equivalent (as much as possible) before the testing begins, and the assumption is that any difference in posttest scores is a result of the treatment. Because all participants have been exposed to the same influences, except for treatment, any different in posttest scores can be attributed, with certain limitations, to the influence of the treatment.

For example, the experimental group consisting of seniors in the graduated care home receive instruction about what and when to eat, and their weight status is evaluated through a posttest. There is no pretest, and the control group receives no treatment.

More questions? See #85, #86, and #87.

QUESTION #81

What Is a One-Shot Case Study, and What Are Some of the Advantages and Disadvantages of Using This Design?

There are many different types of experimental designs, and moving from pre-experimental designs to quasi-experimental designs to true experimental designs, precision increases but also control does as well. This might sound like a desirable set of outcomes, but too much control can lessen generalizability to new settings.

A design with relatively little control but a great deal of generalizability is a one-shot case study, where the group of participants is exposed to a treatment and then the participants are evaluated on some dependent variable(s). For example, a group of middle-aged men and women take a weight training class and are then tested on their skill level.

You may already notice that the most significant shortcoming of this type of design is that there is no control group and also no pretest to help define where the participants started. There is no comparison between these participants and others to find out if there is a difference between groups or a difference over time, so it is very difficult to say whether the weight training had an impact.

However, that said, there are some very good reasons for these one-shot studies.

First, like case studies, they can best be used to study only one participant, institution, or event where the examination can be done in great detail.

Second, a question of group comparison or treatment effects is not an appropriate one to try to answer using this design. Rather, if one is interested in the experience of weight training and what it means to the participants, or how they think they benefited or what they most liked or disliked about it, those might be much more appropriate focal points for the study.

Finally, using this design leads to exploratory research where outcomes suggest further study, such as whether participation in such a program increases social contact, and a more controlled experimental design can be used.

More questions? See #80, #85, and #86.

QUESTION #82

I Know What Correlational Research Methods Are. When and How Are They Used?

You might remember that correlational research methods examine the relationship between two or more variables, such as the number of hours parents spend reading with their children and reading comprehension skills of the child.

What do correlations look like, and what do they mean?

Correlational research relies on (and this is no surprise) correlation coefficients. A correlation coefficient is a numerical index that reflects the association between variables. It ranges from –1.0 to +1.0, such as .75, –.23, or .02 (when positive, the plus sign is usually left out).

In the above reading example, the correlation between these two variables is positive (we know this from previous research), so we can conclude that the more time parents spend reading with the child, the higher the child's reading comprehension score will be. Here, as one variable goes in one direction (more reading by parents), the other variables goes in the same (more reading comprehension by the child).

How about a negative correlation? The more time spent watching TV, the less physically fit an individual is likely to be. The key is that as one variable goes in one direction (more TV), the other variable goes in the opposite (less physically fit).

Here's a summary chart, and ↑ means increases in value and ↓ means decreases in value.

If one variable . . .	And the other . . .	The correlation is called . . .
↑	↑	Positive
↓	↓	Positive
↑	↓	Negative
↓	↑	Negative

Correlation coefficients are based on a grouping of numbers, and if you plotted these values (one dot for each pair), you would get this illustration, where the correlation is .898.

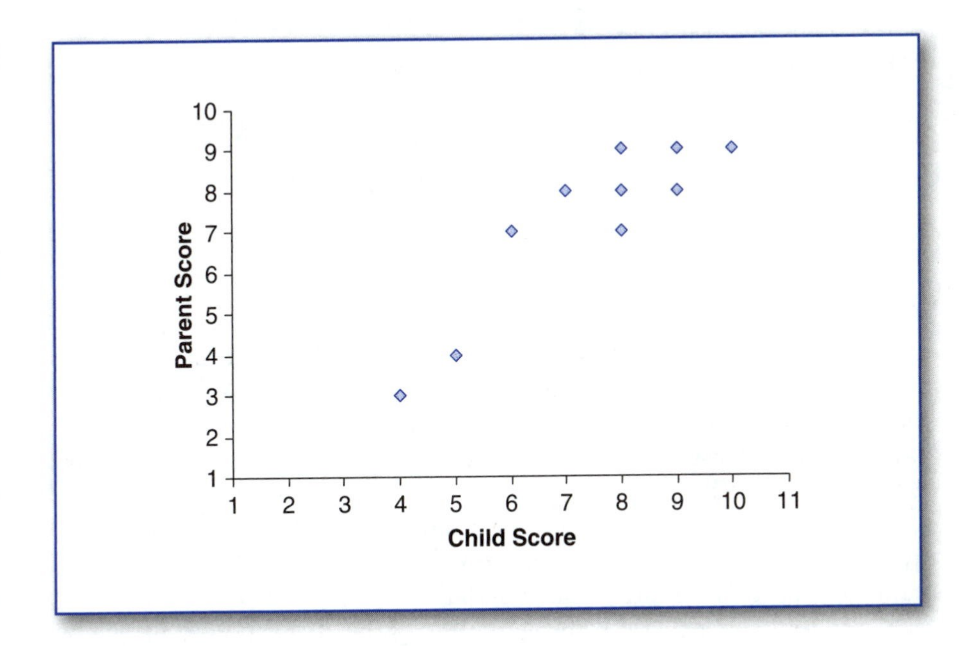

This is a positive correlation, where, as a group and based on these 10 pairs of data points, we can tell that as parent score goes up, so does child score. And, as parent score goes down, so does child score. They move in the same direction—they are positively correlated.

Correlations are represented by a lower case *r* with a subscript of what is being correlated such as $r_{\text{parent children}}$ (with a little dot separating the two).

Primarily, correlations examine the strength of relationships. The one number gives you a very good idea as to how much two things (variables) are related or share in common. The more they share, the stronger the relationship.

But, what they cannot do is provide us with any notion of whether one variable, even if related to the other, has any impact on the value of that variable. In other words, correlations do a very good job of helping us understand the association between variables, but not their causal relationship.

More questions? See #80, #83, and #85.

QUESTION #83

I Know That Correlations Reflect the Association Between Two Variables, but How Do I Interpret Them?

There are basically two ways to interpret a correlation coefficient: one is sort of the eyeball method and the other involves a bit of arithmetic. The eyeball method involves looking at the number and making a decision. For example, if the correlation between motivation and achievement is .7, we judge that (based on previous experience) to be a strong relationship and label it as such.

You remember that a correlation coefficient is a numerical index that reflects the degree of relationship between two variables and that it can range from −1 to +1 and be anything in between (see Question 82). And, you also remember that the absolute value of the correlation coefficient is judging the strength of the correlation and not the sign. So, a correlation of −.4 is stronger than a correlation of +.3.

OK—on to interpretation.

The eyeball method is simply using the following guidelines to judge the strength of this relationship.

If the correlation is between	Then the eyeball method says that this correlation is
.8 and 1.0	Very strong
.6 and .8	Strong
.4 and .6	Moderate
.2 and .4	Weak
0 and .2	Don't even think about it

The much more sophisticated way of judging the strength of a correlation coefficient is by squaring it, which results in what is called r^2. This value is also known as the coefficient of determination, and here's how it works.

A correlation coefficient such as r_{xy} reflects the degree to which x and y are related. But it also reflects how much they share in common with one another—in other words, how much variance each variable shares with one another. r^2 represents the variance in one variable that is accounted for by the variance in the other. The more that is shared, the higher the correlation.

For example, height and weight are usually highly correlated because they share so much of the variance. What's shared? The variables by which height and weight are determined, such as eating habits, genetics, general health, level of exercise, and so forth. Another example is the high correlation we find between socioeconomic status and level of health. What do they have in common? Mostly income. Those people who are in higher income brackets have much easier access to health care and are, in turn, healthier.

So, for example, if the correlation between socioeconomic status and level of health is .7, the coefficient of determination is .49 or 49%, meaning that 49% of the variability in socioeconomic status is shared with level of health.

If the correlation between two variables is 0, they share nothing and have nothing in common. If the correlation is perfect (either +1.0 or –1.0), then the coefficient of determination or r^2 is 100% and they share everything in common. And interestingly, even the eyeball method of evaluating correlations where we deem them to be strong (see above) and between .6 and .8 only find us with coefficients of determination between 36% ($.6^2$) and 64% ($.8^2$)—not particularly high values.

The lesson here? Correlations have to be pretty high for us to have any confidence that the variables involved are meaningfully related.

More questions? See #82, #93, and #94.

QUESTION #84

What Is an Example of a Quasi-Experimental Design, and When Is It Appropriate to Use It?

Quasi-experimental designs are those that contain an independent variable where participants are preassigned to groups. For example, an experiment is conducted to determine if a particular medication improves memory in patients with dementia, and because many such illnesses affect men and women differently, gender is included. There are two independent variables—one is gender and one is level of treatment, such as the dosage of a particular drug. The dependent variable is scores on a test of short-term memory. The design could look like this.

		Dosage Level		
		Dosage 1	**Dosage 2**	**No Treatment**
Gender	**Male**			
	Female			

The gender variable is the one of interest here because its inclusion qualifies this design as quasi-experimental in that participants are preassigned. That is, they are members of the group before the experiment begins, just as with other independent variables such as age, political affiliation, club membership, social class, and income, among many others.

Why is this important for the researcher? Primarily because membership in these groups is not under the researcher's control, and the total precision of a quasi-experimental design is less than what is expected from a true-experimental design, in which all participants are assigned based only on their membership in a treatment group.

The most important use of quasi-experimental designs, and often there is no choice, involves the ethical question of how to study those conditions where participants obviously cannot be assigned to groups, such as an illness, child maltreatment, or divorce. Surely no one would voluntarily contract an illness, or mistreat a child, or get a divorce just to be included in a study, and surely no researcher would create such conditions.

Can you have an experiment, such as the above, where there are "preassigned" independent variables and those that are not? Of course. But, you need to be aware of the difference between the two and the potential limitations that including such "non-treatment" variables creates.

More questions? See #5, #35, and #80.

QUESTION #85

What Is Internal Validity, and Why Is It Important in Experimental Design?

Internal validity is the quality of an experiment where what was manipulated (the independent variable) results in a change in the dependent variable. The notion of internal validity grew out of the seminal work of Donald Campbell and Donald Fiske in their 1963 monograph.

When an experiment is internally valid, then one can point to the results as being a result of the factors that were manipulated in the study. For example, if a study examines the effects of a particular medication on a disease, there will likely be an experimental group that receives the drug and a control group that receives a placebo. All other conditions should be as identical as possible. If the variable of interest is frequency of recurrence of the disease, and if it is shown that only the use of the medication (and no other factor) resulted in a change in the frequency of the disease, one could say that such an experiment has internal validity.

If an experiment does not include a control group, that is obviously one reason why it would not be internally valid—how would one know that the treatment was effective without a comparison? And, other threats to internal validity play an important role as well, such as maturation, history, regression, and confounding.

For example, if only one group of adolescents was used to investigate the impact of interest clubs on the development of friendships during the first year of high school, through the process of emotional, physical, and social growth alone, these young people are sure to change. And, without a comparison group to compare to the group that got the treatment, no sound conclusion might be reached. The obvious solution is to include a control group. This ensures that any difference between the groups (because both mature at the same rate) would be due to participation in interest clubs.

More questions? See #80, #86, and #87.

QUESTION #86

What Is External Validity, and Why Is It Important in Experimental Design?

You already know that internal validity addresses the effectiveness of the independent variable. External validity, another very important term coined by Donald Campbell and Julian Stanley, addresses how generalizable the results of an experiment are. In other words, if an experiment is conducted, how well can the results be generalized to another set of participants in another setting?

Like internal validity, threats to external validity can be easily moderated by the inclusion of a control group, because both groups are similar, if not almost identical, and the only difference observed would be the effects of the treatment.

For example, one threat to external validity can be called *situational effects*. Here, the new setting to which the results are to be generalized may present factors that make the findings less generalizable than they would be otherwise. For example, let's say a study was done to help calm anxious patients who are about to get a flu vaccine, and the treatment (deep breathing exercises) is found to be very effective. However, when it comes time to generalize the results to a new setting, the new setting finds the health care professional coming to the waiting room to call for the next patient holding a hypodermic in his or her hand. The presence alone of the hypodermic could very well affect the participant's level of anxiety regardless of the fact that he or she is about to be inoculated.

The question to ask when regarding external validity is what factors might be present in the new setting that were not present in the original one, that might interfere with the effectiveness of the treatment.

More questions? See #84, #85, and #87.

QUESTION #87

What Is the Trade-Off Between Internal and External Validity?

Although internal and external validity do not have a strong indirect relationship (one increases while the other decreases), there is a bit of a trade-off.

If you remember, a high degree of internal validity is present when the experiment shows that the manipulation of the independent variable results in a change in the dependent variable. A high degree of internal validity is present when the results from the experiment can be generalized to a new setting. Internal validity is all about control (the more control, the more internal validity), whereas external validity is all about the application of findings to new settings.

Here's the question. How can an experiment have sufficient control to adequately test the hypothesis that changes in the independent variable are indeed responsible for changes in the dependent variable and, in fact, generalize to other settings where there is relatively little control? There is no absolute answer to this question and as often stated before in *100 Questions . . .*, the answer depends upon the research question being asked.

If the question is narrow in focus and applies only to some very special conditions where a great degree of control is necessary to adequately test hypotheses, then internal validity is sure to be high. However, it will be tough to make the argument that the experiment has any external validity given how tightly controlled factors are and how hard it is to generalize the findings when there is relatively little control.

If the question is very broad and apparent in the application of the results, it's tough to make the argument that this high degree of external validity comes along with an associated high degree of internal validity because the wide applicability speaks to at least some loss in control of which factors affect which outcomes.

The original question always begs a compromise where the best strategy is to know very well the substance behind the research question being asked and the implications of varying degrees of internal and external validity.

More questions? See #38, #85, and #86.

ALL ABOUT INFERENCE AND SIGNIFICANCE

QUESTION #88

What Is Statistical Significance, and Why Is It Important?

Statistical significance is a measure of the risk that you are willing to take that a null hypothesis is true and you have mistakenly rejected it. The degree of risk you are willing to take is known as significance, or alpha level, and is conventionally set at .05 or .01.

There are two very important and basic foundations to this idea. The first is that when comparisons are made between groups (for example), the assumption is that, having no other knowledge, the two groups are equal to one another. That's our starting point. And, if there is any difference between the two groups, it can be attributed to chance or random variation. So, if Group 1 and Group 2 differ in their attitudes toward advertising, that difference is a function of random variation rather than something specific (such as the coloration of the ad, or the race of the actors in the ad, etc.).

The second is that a comparison is made between the null state (of equality) and the results of the actual experiment, such that the null hypothesis is the most attractive explanation for the lack of a difference, or the research hypothesis is the most attractive explanation for any difference that is observed. If the difference that is observed differs significantly from what we would expect by chance, we speak of that outcome as being significant.

Technically, if the test of the research hypothesis is set at the .05 level of significance, it means that on any one test of the null hypothesis, there is a 5% chance that it will be rejected as not being true when it actually is. And even if the level of significance is set at .0001 (which greatly reduces the chance of making such an error), there is still a chance that such will occur. Nothing's sure, but educated guesses can be somewhat refined through the use of more stringent significance levels.

More questions? See #3, #90, and #96.

QUESTION #89

In Research Reports, I Often See Entries Such as $p = .042$ and $df_{(22)}$. What Do They Mean?

When you read a journal article that tests an empirical hypothesis, you will surely encounter some symbols and abbreviations such as *p* and *df*. There's nothing special about these, but you need to know what they mean to be an informed consumer of research reports.

In research reports, *p* represents probability, and it is more often used when stating the probability of a Type I error. For example, if you see $p = .042$, it means that the probability of making a Type I error is .042. More specifically, it means that on any one test of a true null hypothesis, the likelihood of rejecting it is 4.2 times out of 100. On the next test of that hypothesis (with new data), the *p* value will also change.

When you see *p* values followed by more than (>) and less than (<), which is often the case when the precise value of the Type I error is not available, it is read as it looks . . .

$p < .05$ means that the likelihood of a Type I error is less than 5%

or

$p > .05$ means that the likelihood of a Type I error is more than 5%

When do you see $p < .05$ rather than $p = .042$ (for example)? You usually see the greater and less than signs when a table is used to test the significance of an outcome rather than accessing the exact level of error from the results of a computer analysis. With a table, you can only state that an outcome is more or less likely, but not exactly what the likelihood is.

The term *df* represents degrees of freedom along with the amount (such as 22 in the example $df_{(22)}$). This represents the 22 degrees of freedom with which the statistical outcome is associated, and it is used primarily as one of the variables when looking up the critical value needed for significance.

Both p and df usually occur together in a research report along with the value of the test statistics, so something like

$$t = 1.063, p > .05, df = 16$$

means that a t value of 1.063 with 16 degrees of freedom is not significant beyond the .05 level. You might also see this represented as

$$t_{(16)} = 1.063, p > .05$$

More questions? See #90, #96, and #97.

QUESTION #90

How Do Statistical Programs Such as SPSS Display Significance Levels?

The primary contrast between using a computer program such as SPSS (one of many such programs) and computing significance values manually is ease of use and accuracy.

You may remember that when you determine if an outcome is statistically significant, it involves comparing the obtained level from the analysis to a critical value associated with the type of test that is being used, the significance level at which the research hypothesis is being tested, and the group size. When this comparison is done manually for, say, a simple *t*-test between means, you end up with results such as . . .

$$t_{(22)} = -2.455, p < .05$$

which means that the obtained value of −2.455 is significant at the .05 level with 22 degrees of freedom.

In the case of a computer program, the output and the effort from you is much different.

Here's some selected output from an SPSS program that tests the difference between means as well . . .

Independent Samples Test

		t-test for Equality of Means			
		t	**df**	**Sig. (2-tailed)**	**Mean Difference**
Social Extroversion	Equal variances assumed	−2.455	22	.022	−2.210

Here you can see that the obtained t value is −2.455 and the degrees of freedom are 22, but instead of having to use a table to determine significance, you can see that the *exact* probability of making a Type I error is .022. Much more precise than $p < .05$.

Another reason why computer output might be so attractive is that you can easily cut and paste tables into reports and other types of documents.

More questions? See #88, #89, and #93.

QUESTION #91

What Other Types of Errors Should Be Considered as Part of the Research Process?

When testing a research hypothesis, there are two types of errors that one should consider, and both are the result of the state of the null hypothesis (whether it be true or false) and the action on the part of the research (reject or accept the null hypothesis). Here's a four-way table to help clarify these options.

	Null hypothesis is true	**Null hypothesis is false**
Reject the null hypothesis	a. Type I error or alpha (α) or level of significance	b. ☺ Correct decision
Accept the null hypothesis	c. ☺ Correct decision	d. Type II error or beta (β)

And, as an example, let's assume that the null hypothesis is that two groups of middle-aged factory workers really do not differ in their scores on a recall of past events test. Conditions b and c above would be correct if, based on the results of the data and their analysis, you accept the null if it is true and you reject it if it is false. Remember that because the population is never tested directly, you never really know whether the null is "truly" true or false.

Under Condition a, you would be making a Type I error because you reject the null when it is true. For example, there really is no difference between the two groups but you conclude that the Type 1 error defines the probability of making such an error on every test of the null hypothesis.

The second kind of error can be made when the null hypothesis is false (there really is a difference between the two groups), but you mistakenly accept the null hypothesis that there is not a difference.

In both of these cases of error, you can make some adjustments to minimize the mistake. First, you set the alpha level and test the research hypothesis at that chosen value. Second, the larger the sample, the more likely it is that the level of Type II error will be reduced.

More questions? See #88, #92, and #96.

QUESTION #92

What Is Power, and Why Is It Important?

You may remember that a Type I error is also known as level of significance or alpha or α and is the rejection of a true null hypothesis. We want to avoid those kinds of errors. Another type of error we want to avoid occurs when you accept a false null hypothesis. Committing such an error is known as a Type II error or beta or β.

Power is the ability to detect and reject a false null hypothesis and is equal to $1 - \beta$. It is the combination of the sample size and the technique being used that will allow us to minimize Type II error, where something in fact is not true (the null hypothesis) yet we fail to reject it.

Power is a quality of a test of a research hypothesis that we try to assess prior to the actual test of that research hypothesis, and even before in the design of the experiment, and it usually consists of three different factors:

- How large we expect the effect of the treatment to be
- The size of the sample
- The type I error rate we set.

The larger the sample (or sample size), the easier it will be to find a significant difference and the more power the test has to detect and reject false null hypotheses.

As the sample size increases, the sample is more representative of the population and better mimics the characteristics of the population. Hence, one needs a smaller sample than usual to detect any differences.

Finally, as the level of significance becomes less conservative (from, say, .05 to .01), it is "easier" to find significance because the critical value needed to not accept the null is lower.

Any of these three conditions (magnitude of the difference, sample size, and Type I error) can have an impact on power, but they can also each be isolated and manipulated to help increase the odds that the test of any one research hypothesis is powerful enough to detect false differences.

More questions? See #88, #91, and #96.

QUESTION #93

What Are Some of the Other Popular Statistical Tests, and When Are They Used?

There are hundreds of statistical tests, and before we review some of the most often used, there are a few guidelines you need to follow when selecting any test.

- The test you select depends upon the research question that you are asking and the method you use to answer that question and test the associated hypothesis. An often-made mistake occurs when beginning researchers do not collect the type of data needed to answer the question they are asking and are unable to select the correct statistical test.
- All statistical tests result in an obtained value. You then have to judge whether that value is significant or not. You can do this through the use of tables (usually in the back of introductory statistical textbooks) and through the use of statistical analysis programs.
- Statistical tests are tools. You are best served to understand how to compute the resulting values by hand before you perform these tests using a computer. In some cases (such as factor analysis), the techniques may be too complex, but in most others, manual computation is still the best way to learn.

Here is a brief summary of some of the most frequently used statistical tests (those that appear most often in behavioral and social science journals) and each does.

If you want to...	**Use this statistical technique...**
Test for the difference between the means of two groups	*t*-test
Test for differences between the means of two or more groups on one factor	One-way analysis of variance
Tests for differences between the means of two or more groups on more than one factor	Two-way analysis of variance
Test for the significance of a regression coefficient on how well one or more variables predict an outcome	*F*-test
Test for the presence of underlying factors	Factor analysis
Compare one group to some hypothesized outcome	Z test
Find out how well one or more variables predict one or more other variables	Regression

More questions? See #89, #95, and #97.

QUESTION #94

What Is Regression, and How Is It Used?

Regression is a very powerful technique that answers the question of how well one or more variable(s) predict another variable. In regression, the predictor variable (X) is known as the independent variable and the predicted variable (Y) is known as the dependent variable. For example, if you were interested in knowing how well senior class rank in high school predicts first-semester college grades, you could do the following.

1. For a sample of high school seniors, collect their rankings.
2. For the same sample, once they get to college, collect their first-semester grade point average.
3. A regression equation such as the following is then created and used to predict an outcome.

$$\text{Predicted GPA } (Y) = \text{Senior Class Rank } (X)$$

Now, using a simple regression question, take rankings for those seniors not yet in college and predict their GPAs from the ranking. The stronger the correlation between rank and GPA, the better the prediction. If the correlation were perfect, then one could perfectly predict GPA. But, because correlations are almost never perfect or $+/-1.0$, regression is a good, but not perfect, solution. Just as there are always possibilities that we are incorrect, such is the case with regression as well. A score can be predicted, but the error of estimate is the distance an actual score is from a predicted score. The better the prediction, the lower the error of estimate and the more accurate the model.

A more sophisticated and often used technique is multiple regression. Here, more than one factor (or X variable) is used to predict an outcome or a Y variable. For example, in addition to class ranking being used as a predictor of GPA, we can add hours of community service and number of extracurricular activities in which seniors participated and do the same thing.

Rather than the simple equation you saw earlier, it's expanded as such . . .

Predicted GPA (Y) = Senior Class Rank (X_1) + Hours of Community Service (X_2) + Hours of Curricular Activities (X_3)

And you will not be surprised to learn that the more predictors used, the better the prediction. But, too many, and their use can get redundant, costly, and inefficient.

More questions? See #83, #88, and #93.

QUESTION #95

What Is the Difference Between a Parametric and a Nonparametric Test?

Both parametric and nonparametric tests are tests of inference. There are several differences between the two that determine which is appropriate to use when. Here are some of the most important guidelines, but keep in mind that special circumstances always exist where you may use one or the other type of test.

1. Parametric tests assume a normal distribution, whereas nonparametric tests assume any type of distribution.
2. Parametric tests assume that variances within groups are equal, whereas nonparametric tests assume they are not.
3. The type of data that is analyzed using parametric tests has an interval or ratio level of measurement, whereas with nonparametric tests, the level of measurement is usually nominal or ordinal.
4. Parametric tests are usually done on samples that are larger than 30, whereas nonparametric tests are done on smaller groups.

And, for the several common types of questions we ask as researchers, there are corresponding parametric and nonparametric tests. For example, here are the questions asked and the corresponding parametric and nonparametric tests.

Test Name	When to Use It
McNemar Test for Significance of Changes	Examining before-and-after changes
Fisher's Exact Test	Computing the exact probability of an outcome
Chi-Square One-Sample	Determining if the number of occurrences of an event is random
Sign Test	Comparing the medians from two samples
Mann-Whitney U Test	Comparing two independent samples
Friedman Two-Way Analysis of Variance	Examining differences between two or more independent samples

One more important caveat—nonparametric tests are not just a convenience to be used when the sample is not normal or when it is more convenient to keep a sample relatively small. They have very definite applications and, as with all research methods, need to be used judiciously.

More questions? See #88, #91, and #97.

QUESTION #96

I Often See the Term "Statistical Significance" Being Used in Journal Articles. What Is It, and Why Is It Important?

An important part of the scientific method is to test hypotheses. And, researchers are always faced with the question of what criterion to use to conclude that the outcomes of such a test are of value and, as best as can be determined, "truthful."

To answer this question, the concept of statistical significance is used. Statistical significance is the likelihood that even if the results of an analysis are in the direction predicted by the hypothesis, there is always a chance that the wrong conclusion is being reached. The probability of making a mistake like this is also called the significance level.

You remember that there are two basic types of hypotheses—a null hypothesis and a research hypothesis. A null hypothesis is a statement of equality, and a research hypothesis is a statement of inequality. Even though the null hypothesis cannot be tested directly (because we test samples, and null hypotheses are relevant only to populations), we look to it when defining the concept of significance.

Technically, statistical significance is defined as the likelihood of rejecting a null hypothesis when it is actually true. For example, our results might show that two groups are not equal to one another, but in truth, they are equal. The likelihood of making that mistake reflects the concept of statistical significance.

Statistical significance is the likelihood that such an error is made. Here, the researcher is confident in the results of the experiment being accurate, but because the research process is not perfect, there is always a small chance (and that's exactly what significance level is—the probability of an error being made) that he or she will reach the wrong conclusion.

How big might these mistakes be? Conventionally, the level of statistical significance is reported as .01 or .05. Consistent with the above comments, a significance level of .01 means that there is a 1% chance on the test of any null hypothesis that it will be rejected when it is indeed true.

More questions? See #8, #89, and #91.

QUESTION #97

How Can I Tell If an Outcome Is Statistically Significant?

Every statistical test has a particular distribution for that test statistic associated with it. So, for example, the *t*-test for differences between means has associated with it a distribution of all possible values of the *t* statistic. This is the case for every inferential test that you will learn about and use in your research. Each of these distributions also has a table associated with it; a portion of the table for *t*-tests is shown below.

The important elements of this table are degrees of freedom roughly corresponding to sample size; a one- or two-tailed test depending upon whether the hypothesis is directional (one-tailed) or nondirectional (two-tailed); the level of significance at which the research hypothesis is being tested; and, finally, the critical *t* values. The tabled critical value represents the point in the distribution beyond which the test outcome is so unlikely that it is almost impossible to attribute it to chance.

Degrees of Freedom	One-Tailed Test		Two-Tailed Test	
	.05	.01	.05	.01
20	1.725	2.528	2.086	2.846
21	1.721	2.518	2.080	2.832
22	1.717	2.509	2.074	2.819
23	1.741	2.500	2.069	2.808

Each statistical test produces an obtained value, such as in the case of a hypothetical test between means where $t = 1.834$, the degrees of freedom (or *df*) are 21, and the directional or one-tailed hypothesis is being tested at the .05 value. The obtained value (in this hypothetical example, 1.834) is compared to the table's critical value, which in this case is 1.721. If the critical value exceeds the obtained value, the null hypothesis is the most attractive explanation for any observed group difference. If the obtained value exceeds the critical value, the research hypothesis is the most attractive explanation for any observed group difference.

Also, most statistical analysis programs produce a specific Type I error or level of significance value. They all differ in how they look, but most will produce something that looks like this . . .

$$p = .034$$

which is the exact probability of the outcome described above occurring by chance alone.

More questions? See #80, #88, and #100.

QUESTION #98

What Is Effect Size?

Effect size is a measure of how much two groups differ from one another, but more technically, it is a measure of the magnitude of a treatment's effect. In conjunction with the results of a statistical test (what the Type I error or level of significance might be), it provides additional information as to whether the treatment worked.

The most simple way to compute effect size is by using the following formula for computing effect size for a test of differences between two averages. By examining its components, you can get a very good idea as to how it works.

$$ES = \frac{\bar{X}_1 - \bar{X}_2}{s}$$

where *ES* equals effect size, $\bar{X}$ equals the means of Group 1 and Group 2, and *s* is the standard deviation from either group.

It should be obvious that as the difference between the groups' averages gets larger, the effect size goes up. And, as the variability within either of the groups increases, the effect size goes down. Both of these make sense. As the difference between the groups increases, the effect of the treatment is likely to increase. And, as the variability within either group increases, the differences within the group increase and the treatment is less likely to have a large effect.

Once you have this number, how do you interpret it? As with so many things regarding research methods in the social and behavioral sciences, there's a bit of art and a bit of science to it, but you can use the following guidelines as a starting point.

- A small effect size ranges from 0 to .20
- A moderate effect size ranges from .2 to .50
- A large effect size exceeds .50

For example, an effect size of .74 tells us that two groups are far apart from each other and overlap little. Likewise, a small effect size of, say, .13 tells us that the two groups are very similar with a great deal of overlap. Effect size used in conjunction with significance level should provide a very clear picture of the importance and meaningfulness of the difference between two groups.

More questions? See #80, #88, and #91.

QUESTION #99

What Is the Difference Between Statistical Significance and Meaningfulness?

Statistical significance is the probability assigned to a particular outcome and is shown as "beyond the .05 level" (shown as $p < .05$) or "beyond the .01 level" (shown as $p < .01$). Also, specific probability can be assigned to outcomes such as $p = .238$ or $p = .011$.

Statistical significance is a very useful construct and one that is used quite often in research activities. It is often mistakenly held up as the most important criterion to determine the usefulness of experimental findings.

However, regardless of how useful a tool it is, a very important dimension of any research study and accompanying statistical analysis is left out—the evaluation of the finding's meaningfulness. This distinction between statistical significance and meaningfulness can best be shown through an example.

Imagine a large-scale study costing $400,000 that is examining the effects of an early intervention program enrolling 500 developmentally delayed toddlers on their later social skills. One group of infants receives the treatment and the other does not, and the outcome variable is a measure of social skills. An analysis of the data reveals that, indeed, the social skills of those children involved in the study are higher and significantly different from those who did not receive any treatment. The experimental group scored 87.4 out of a possible 100 points, and the control group scored 87.2.

What should one make of these results, and how does one evaluate them fully from more than just the perspective of statistical significance? Although the difference is statistically significant, does a .2-point increase in scores result in such a meaningful increase that the $400,000 cost of the program can be justified? Maybe yes and maybe no, but statistical significance is not enough. The cultural or contextual value of the outcomes, what we call meaningfulness, has to be considered as well.

More questions? See #8, #88, and #92.

QUESTION #100

Why Are the Values of .01 and .05 Usually Used as Conventional Levels of Statistical Significance?

This is a little bit of a trick question—important to answer, but it has a much better lesson at the end. First, a little history adapted from a paper by Michael Cowles and Caroline Davis from York University and published in the *American Psychologist.*

The notion of .05 as a conventional Type I error rate seems to have gotten its start in the early work of R. A. Fisher, the British statistician who created analysis of variance as a method for looking at differences between more than two means. Fisher thought that an outcome that had a probability associated with it, such as 1 in 20 (5%), was sufficient to see the outcome as "significant."

However, lots went on before Fisher's comments and direction. A great deal of work in the mathematics of probability started in 17th-century France, where insurance actuarials estimated the likelihood of someone being a "lunatik" (mentally ill) and associated probabilities with that possible outcome. From there, additional mathematicians got into the game, eventually turning to the normal curve to estimate the likelihood of non-chance occurrence for everything from soldiers' height to other physical characteristics—once again, looking for the cut point where outcomes exceeded "normality."

Since that time, probable error and critical ratios have been used instead, and later statisticians such as Fisher and Karl Pearson continued to lead the way in attempting to define a level of chance that most reasonable people would consider representative of a rare event. The value of .05 is definitely subjective, and it is not based on any long and technical mathematical or conceptual argument. However, it *is* based on the early work of many scientists who looked at particular real-life outcomes and decided that this cut point seems to accommodate most estimates of what is a function of chance versus what is a function of outset factors such as experimental treatments. Remember that most analyses completed today specify the exact level of a Type I error, and there's no need for a margin of error such as $p < .5$.

More questions? See #8, #88, and #91.

Index